Little Brothers of the Air

OLIVE THORNE MILLER

Idle Winter Press
Portland, Oregon

Interior text content by Harriet Mann Miller (Olive Thorne Miller) was originally published in 1893 by Houghton, Mifflin & Co., The Riverside Press, Cambridge Massachusetts, U.S.A.; electrotyped and printed by H. O. Houghton & Co. and is in the public domain. Bird species common and scientific names have been modernized in the index.

Idle Winter Press
Portland, Oregon
http://IdleWinter.com

First published 1893
This edition published 2015
Printed in the United States of America

The text of this book is in Alegreya

ISBN-13: 978-0692381229 (Idle Winter Press)
ISBN-10: 0692381228

INTRODUCTORY

Some of the chapters of this little book were written in 1888, on the shore of the Great South Bay, Long Island; others in the northern part of New York State, known to its residents as the "Black River Country," a year or two later. Part of them have been published in The Atlantic Monthly, Harper's Bazaar, The Independent and other papers.

The nomenclature featured in the Table of Contents is that adopted by the American Ornithological Society [as of the original publication in 1892; nomenclature in the Index has been updated as of 2015].

Olive Thorne Miller

CONTENTS

ON THE GREAT SOUTH BAY

IN THE BLACK RIVER COUNTRY

ON THE GREAT SOUTH BAY

Precious qualities of silence haunt
Round these vast margins ministrant.

'T is here, 't is here, thou canst unhand thy heart
And breathe it free, and breathe it free
By rangy marsh, in lone sea-liberty.

Sidney Lanier

I. THE KINGBIRD'S NEST

To study a nest is to make an acquaintance. However familiar the bird, unless the student has watched its ways during the only domestic period of its life—nesting time—he has still something to learn. In fact, he has almost everything to learn, for into those few weeks is crowded a whole lifetime of emotions and experiences which fully bring out the individuality of the bird. Family life is a test of character, no less in the nest than in the house. Moreover, to a devotee of the science that some one has aptly called Ornithography, nothing is so attractive. What hopes it holds out! Who can guess what mysteries shall be disclosed, what interesting episodes of life shall be seen about that charmed spot?

To find a newly built nest is the first June work of the bird-student, and this year on the Great South Bay a particularly inviting one presented itself, on the top branch of a tall oak-tree near my "inn of rest." It was in plain sight from the veranda. The builder evidently cared nothing for concealment, and relied, with reason, upon its inaccessible position for safety. To be sure, as days went by and oak leaves grew, a fair screen for the little dwelling was not lacking; but summer breezes were kind, and often blew them aside, and, better still, from other points of view the nest was never hidden.

To whom, then, did the nest belong? I hoped to the kingbird, who at that moment sat demurely upon the picket fence below, apparently interested only in passing insects; and while I looked the question was answered by Madame Tyrannis herself, who came with the confidence of ownership, carrying a beakful of building material, and arranging it with great pains inside the structure. This was satisfactory, for I did not know the kingbird in domestic life.

For several days it seemed uncertain whether the kingbirds would ever really occupy the nest, so spasmodic was the work upon it. Now one of the pair came with a bit of something, placed it, tried its effect this way and that, and then disappeared; while for hours every day both might be seen about the place, hunting insects and taking their ease on the fence as if no thought of nesting ever stirred their wise little heads. The last addition to the domicile was curious: a soft white feather from the poultry yard, which was fastened up on the edge, and stood there floating in the breeze; a white banner of peace flung out to the world from her castle walls.

Peace from a kingbird? Direful tales are told of this bird: "he is pugnacious," says one writer; "he fights everybody," adds another; "he is a coward," remarks a third. Science has dubbed him tyrant (*Tyrannis*), and his character is supposed to be settled. But may there not be two sides to the story? We shall see. One kingbird, at least, shall be studied sympathetically; we shall try to enter his life, to judge him fairly, and shall above all

"bring not
The fancies found in books,
Leave author's eyes, and fetch our own."

Nearly two months that small dwelling on the oak was watched, day after day, early and late, in storm and in sunshine; now I know at least one family of kingbirds, and what I know I shall honestly tell, "nothing extenuating."

The house was built, the season was passing, yet housekeeping did not begin. The birds, indeed, appeared to have abandoned the tree, and days went by in which I could not see that either visited it. But the nest was not deserted, for all that; the curiosity and impertinence of the avian neighbors were simply amazing. (Perhaps the kingbird has some reason to be pugnacious!) No sooner was that tenement finished than, as promptly as if they had received cards to a house-warming, visitors began to come. First to show himself was an orchard oriole, who was in the habit of passing over the yard every day and stopping an hour or more in the neighborhood, while he scrambled over the trees, varying his lunches with a rich and graceful song. Arrived this morning in the kingbird tree, he began his usual hunt over the top branch, when suddenly his eye fell upon the kingbird cradle. He paused, cast a wary glance about, then dropped to a lower perch, his singing ended, his manner guilty.

Nearer and nearer he drew, looking cautiously about and moving in perfect silence. Still the owner did not come, and at last the stranger stood upon the edge. What joy! He looked that mansion over from foundation to banner fluttering in the wind; he examined closely its construction; with head turned over one side, he criticized its general effect, and apparently did not think much of it; he gratified to the full his curiosity, and after about one minute's study flew to the next tree, and resumed his singing.

The next arrival was a pewee, whose own nest was nearly built, in a wild-cherry tree not far off. The fence under the oak was his usual perch, and it was plain that he made his first call with "malice aforethought;" for, disdaining the smallest pretense of interest in it, he flew directly to the nest, hovered beneath it, and pulled out some part of the building material that pleased his fancy—nothing less than pure thievery.

Among the occasional visitors to the yard were two American goldfinches, or thistle-birds, in bright yellow and black plumage, both males. They also went to the new homestead in the oak, inspected it, chatted over it in their sweet tones, and then passed on. It began to look as though the nest were in the market for any one to choose, and the string of company was not yet ended.

Soon after the goldfinches had passed by, there alighted a gay Baltimore oriole, who, not content with looking at the new castle in the air, must needs try it. He actually stepped into the nest and settled down as if sitting. Who knows but he was experimenting to see if this simple, wide-open cradle wouldn't do as well for oriole babies as it did for kingbirds? Certainly it was a curious performance. It made an impression on him too, for the next day he came again; and this time he picked at it, and seemed to be changing its interior arrangement, but he carried nothing away when he flew.

Even after sitting began, this oriole paid two more visits to the nest which so interested him. On the first occasion, the owner was at home, and gave him instant notice that the place was no longer on view. He retired, but, being no coward, and not choosing to submit to dictation, he came again. This time, a fly-up together, a clinch in the air, with loud and offensive remarks, cured him of further desire to call.

More persistent than any yet mentioned was a robin. Heretofore, strange to say, the guests had all been males, but this caller was the mother of a young brood in the next yard. She came in her usual way, alighted on a low branch, ran out upon it, hopped to the next higher, and so proceeded till she reached the nest. The kingbird happened to be near it himself, and drove her away in an indifferent manner, as if this interloper were of small account. The robin went, of course, but returned, and, perching close to the object of interest, leaned over and looked at it as long as she chose, while the owner stood calmly by on a twig and did not interfere.

I know he was not afraid of the robin, as later events proved; and it really looked as if the pair deliberately delayed sitting to give the neighborhood a chance to satisfy its curiosity; as if they thus proclaimed to whom it might concern that there was to be a kingbird household, that they might view it at their leisure before it was occupied, but after that no guests were desired. Whatever the cause, the fact is, that once completed, the nest was almost entirely abandoned by the builders for several days, during which this neighborhood inspection went on. They even deserted their usual hunting-ground, and might generally be seen at the back of the house, awaiting their prey in the most unconcerned manner.

However, time was passing; one day Madame Tyrannis herself began to call, but fitfully. Sometimes she stayed about the nest one minute, sometimes five minutes, but was restless; picking at the walls, twitching the leaves that hung too near, rearranging the lining, trying it this way and that, as if to see how it fitted her figure, and how she should like it when she was settled. First she tried sitting with face looking toward the bay; then she jerked herself around, without rising, and looked awhile toward the house. She had as much trouble to get matters adjusted to her mind as if she had a houseful of furniture to place, with carpets to lay, curtains to hang, and the thousand and one "things" with which we bigger housekeepers cumber ourselves and make life a burden. This spasmodic visitation went on for days, and finally it was plain that sitting had begun. Still the birds of the vicinity were interested callers, and I began to think that one kingbird would not even protect his nest, far less justify his reputation by tyrannizing over the feathered world. But when his mate had seriously established herself, it was time for the head of the household to assume her defense, and he did.

As usual, the kingbird united the characters of brave defender and tender lover. To his spouse his manners were charming. When he came to relieve her of her care, to give her exercise or a chance for luncheon, he greeted her with a few low notes, and alighted on a small leafless twig that curved up about a foot above the nest, and made a perfect watch-tower. She slipped off her seat and disappeared for about six minutes. During her absence he stayed at his post, sometimes changing his perch to one or other of half a dozen leafless branchlets in that local part of the tree, and there sitting, silent and watchful, ready to interview any stranger who appeared.

Upon her return he again saluted her with a few words, adding to them a lifting of wings and spreading of his beautiful tail that most comically suggested the bowing and hat-lifting of bigger gentlemen. In all their life together, even when the demands of three infants kept them busy from morning till night, he never forgot this little civility to his helpmate. If she alighted beside him on the fence, he rose a few inches above his perch, and flew around in a small circle while greeting her; and sometimes, on her return to the nest, he described a larger circle, talking (as I must call it) all the time.

Occasionally, when she approached, he flew out to meet and come back with her, as if to escort her. Could this bird, to his mate so thoughtful and polite, be to the rest of the world the bully he is pictured? Did he, who for ten months of the year shows less curiosity about others, and attends more perfectly to his own business than any bird I have noticed, suddenly, at this crisis in his life, become aggressive, and during these two months of love and paternity and hard work, make war upon a peaceful neighborhood?

I watched closely. There was not an hour of the day, often from four a.m. to eight p.m., that I had not the kingbird and his nest directly in sight, and hardly a movement of his life escaped me. There he stood, on the fence under his tree, on a dead bush at the edge of the bay, or on the lowest limb of a small pear-tree in the yard. Sometimes he dashed into the air for his prey; sometimes he dropped to the ground to secure it; but oftenest, especially when baby throats grew clamorous, he hovered over the rank grass on the low land of the shore, wings beating, tail wide spread, diving now and then for an instant to snatch a morsel; and every thirty minutes, as punctually as if he carried a watch in his trim white vest, he took a direct line for the home where his mate sat waiting.

A few days after the little dame took possession of the nest, the kingbird had succeeded, without much trouble, in making most of his fellow-creatures understand that he laid claim to the upper branches of the oak, and was prepared to defend them against all comers, and they simply gave the tree a wide berth in passing.

Apparently deceived by his former indifference, however, the robin above mentioned presumed to call somewhat later. This time she was received in a manner that plainly showed she was no longer welcome. She retired, but she expressed her mind freely for some time, sitting on the fence below. With true robin persistence she did not give it up, and she selected for her next call the dusk of evening, just before going to bed.

This time both kingbirds flung themselves after her, and she left, "laughing" as she went. The kingbirds did not follow beyond their own borders, and the robin soon returned to the nearest tree, where she kept up the taunting "he! he! he!" a long time, seemingly with deliberate intention to insult or enrage her pursuers, but without success; for unless she came to their tree, the kingbirds paid her not the slightest attention.

The last time the robin tried to be on friendly terms with her neighbor, I noticed her standing near him on the picket fence under his tree. There were not more than three pickets between them, and she was expostulating earnestly, with flirting tail and jerking wings, and with loud "tut! tut's," and "he! he's!" she managed to be very eloquent. Had he driven her from his nest? and was she complaining? I could only guess. The kingbird did not reply to her, but when she flew he followed, and she did not cease telling him what she thought of him as she flew, till out of sight.

Strangest of all was the fact that, during the whole of this scene, her mate stood on the fence within a dozen feet, and looked on! Did he think her capable of managing her own affairs? Did he prefer to be on good terms with his peppery neighbor? or was it because with her it would be a war of words, while if he entered the arena it must be a fight? as we sometimes see, when a man goes home fighting drunk, every man of the neighborhood keeps out of sight, while all the women go out and help his wife to get him home.

The most troublesome meddler was, as might be expected, an English sparrow. From the time when the first stick was laid till the babies were grown and had left the tree, that bird never ceased to intrude and annoy. He visited the nest when empty; he managed to have frequent peeps at the young; and notwithstanding he was driven off every time, he still hung around, with prying ways so exasperating that he well deserved a thrashing, and I wonder why he did not get it. He was driven away repeatedly, and he was "picked off" from below, and pounced upon from above, but he never failed to return.

Another visitor of whom the kingbird seemed suspicious was a purple crow blackbird, who every day passed over. This bird and the common crow were the only ones he drove away without waiting for them to alight; and if half that is told of them be true, he had reason to do so.

With none of these intruders had the kingbird any quarrel when away from his nest. The blackbird, to whom he showed the most violence, hunted peacefully beside him on the grass all day; the robin alighted near him on the fence, as usual; the orioles scrambled over the neighboring trees, singing, and eating, as was their custom; even the English sparrow carried on his vulgar squabbles on his own branch of the oak all day; but to none of them did the kingbird pay the slightest attention. He simply and solely defended his own household.

In the beginning the little dame took sitting very easy, fidgeting about in the nest, standing up to dress her feathers, stretching her neck to see what went on in the yard below, and stepping out upon a neighboring twig to rest herself. After a few days she settled more seriously to work, and became very quiet and patient.

Her mate never brought food to her, nor did he once take her place in the nest; not even during a furious northeast gale that turned June into November, and lasted thirty-six hours, most of the time with heavy rain, when the top branch bent and tossed, and threatened every moment a catastrophe. In the house, fires were built and books and work brought out; but the bird-student, wrapped in heavy shawls, kept close watch from an open window, and noted ardently the bad-weather manners of Tyrannis.

Madame sat very close, head to the northeast, and her tail, narrowed to the width of one feather, pressed against a twig that grew up behind the nest. All through the storm, I think the head of the family remained in a sheltered part of the tree, but he did not come to the usual twigs which were so exposed. I know he was near, for I heard him, and occasionally saw him standing with body horizontal instead of upright, as usual, the better to maintain his position against the wind. At about the ordinary intervals the sitter left her nest, without so much as a leaf to cover it, and was absent perhaps half as long as common, not once did her mate assume her post.

How were this pair distinguished from each other, since there is no difference in their dress? First, by a fortunate peculiarity of marking; the male had one short tail feather, that, when he was resting, showed its white tip above the others, and made a perfectly distinct and (with a glass) plainly visible mark. Later, when I had become familiar with the very different manners of the pair, I did not need this mark to distinguish the male, though it remained *en evidence* all through the two months I had them under observation.

During the period of sitting, life went on with great regularity. The protector of the nest perched every night in a poplar-tree across the yard, and promptly at half past four o'clock every morning began his matins. Surprised and interested by an unfamiliar song, I rose one day at that unnatural hour to trace it home. It was in that enchanting time when men are still asleep in their nests, and even "My Lord Sun" has not arisen from his;

when the air is sweet and fresh, and as free from the dust of man's coming and going as if his tumults did not exist. It was so still that the flit of a wing was almost startling. The water lapped softly against the shore; but who can

> "Write in a hook the morning's prime,
> Or match with words that tender sky"?

The song that had called me up was a sweet though simple strain, and it was repeated every morning while his mate was separated from him by her nest duties. I can find no mention of it in books, but I had many opportunities to study it, and thus it was. It began with a low kingbird "Kr-r-r" (or rolling sound impossible to express by letters), without which I should not have identified it at first, and it ended with a very sweet call of two notes, five tones apart, the lower first, after a manner suggestive of the phœbe—something like this: "Kr-r-r-r-r-ree-bé! Kr-r-r-r-ree-bé!" In the outset, and I think I heard the very first attempt, it resembled the initial efforts of cage-birds, when spring tunes their throats. The notes seemed hard to get out; they were weak, uncertain, fluttering, as if the singer were practicing something quite new. But as the days went by they grew strong and assured, and at last were a joyous and loud morning greeting. I don't know why I should be so surprised to hear a kingbird sing; for I believe that one of the things we shall discover, when we begin to study birds alive instead of dead, is that every one has a song, at least in spring, when, in the words of an enthusiastic bird-lover, "the smallest become poets, often sublime songsters." I have already heard several sing that are set down as lacking in that mode of expression.

To return to my kingbird, struggling with his early song. After practicing perhaps fifteen or twenty minutes, he left his perch, flew across the yard, and circled around the top bough, with his usual good-morning to his partner, who at once slipped off and went for her breakfast, while he stayed to watch the nest.

This magic dawn could not last. It grew lighter; the sun was bestirring himself. I heard oars on the bay; and now that the sounds of men began, the robin mounted the fence and sang his waking song. The rogue! —he had been "laughing" and shouting for an hour. "Awake! awake!" he seemed to say; and on our dreamy beds we hear him, and think it the first sound of the new day. Then, too, came the jubilee of the English sparrow, welcoming the appearance of mankind, whose waste and improvidence supply so easily his larder. Why should he spend his time hunting insects? The kitchen will open, the dining-room follows, and crumbs are sure to result. He will wait, and meanwhile do his best to waken his purveyor.

I found this to be the almost invariable programme of kingbird life at this period: after matins, the singer flew to the nest tree, and his spouse went to her breakfast; in a few seconds he dropped to the edge of the nest, looked long and earnestly at the contents, then flew to one of his usual perching-places near by, and remained in silence till he saw the little mother coming. During the day he relieved her at the intervals mentioned, and at night, when she had settled to rest, he stayed at his post on the fence till almost too dark to be seen, and then took his way, with a good-night greeting, to his sleeping-place on the poplar.

Thus matters went through June till the 29th, when, at about four o'clock in the afternoon, there was an unusual stir about the kingbird castle. I did see that something had happened, and this must open a new chapter. But before beginning the next chronicle of the kingbird babies, I should like to give my testimony about *one* member of the family. As a courteous and tender spouse, as a devoted father and a brave defender of his household, I know no one who outranks him. In attending to his own business and never meddling with others, he is unexcelled.

In regard to his fighting, he has driven many away from his tree, as do all birds, but he never sought a quarrel; and the only cases of anything like a personal encounter were with the two birds who insisted on annoying him. He is chivalrous to young birds not his own, as will appear in the story of his family. He is, indeed, usually silent, perhaps even solemn, but he may well be so; he has an important duty to perform in the world, and one that should bring him thanks and protection instead of scorn and a bad name. It is to reduce the number of man's worst enemies, the vast army of insects. What we owe to the fly-catchers, indeed, we can never guess, although, if we go on destroying them, we may have our eyes opened most thoroughly. Even if the most serious charge against the kingbird is true, that he eats bees, it were better that every bee on the face of the earth should perish than that his efficient work among other insects should be stopped.

II. A CHRONICLE OF THREE LITTLE KINGS

There was a

> "Riot of roses and babel of birds,
> All the world in a whirl of delight,"

when the three baby kingbirds opened their eyes to the June sunlight. Three weeks I had watched, if I had not assisted at, the rocking of their cradle, followed day by day the patient brooding, and carefully noted the manners and customs of the owners thereof. At last my long vigil was rewarded. It was near the end of a lovely June day, when June days were nearly over, that there appeared a gentle excitement in the kingbird family. The faithful sitter arose, with a peculiar cry that brought her mate at once to her side, and both looked eagerly together into the nest that held their hopes. Once or twice the little dame leaned over and made some arrangements within, and then suddenly she slipped back into her place, and her spouse flew away. But something had happened, it was plain to see; for from that moment she did not sit so closely, her mate showed unusual interest in the nest, and both of them often stood upon the edge at the same time. That day was doubtless the birthday of the first little king.

To be sure, the careful mother still sat on the nest part of every day, but that she continued to do, with ever-lengthening intervals, till every infant had grown up and left the homestead forever.

All through the sitting the work of the head of the family had been confined to encouraging his partner with an early morning song and his cheerful presence during the day, and to guarding the nest while she sought her food; but now that her most fatiguing labor was over, his began. At first he took entire charge of the provision supply, while she kept her nurslings warm and quiet, which every mother, little or big, knows is of great importance. When the young father arrived with food, which he did frequently, his spouse stepped to the nearest twig and looked on with interest, while he leaned over and filled one little mouth, or at any rate administered one significant poke which must be thus interpreted. He did not stay long; indeed, he had not time, for this way of supplying the needs of a family is slow business; and although there were but three mouths to fill, three excursions and three hunts were required to fill them.

In the early morning he seemed to have more leisure; at that time, the happy young couple stood one each side of the nest, and the silent listener would hear the gentle murmurs of what Victor Hugo calls "the airy dialogues of the nest." Ah, that our dull ears could understand!

For some days the homestead was never left alone, and the summer breezes

"Softly rocked the babies three,
Nestled under the mother's wing,"

almost as closely as before they came out of the egg. But much of the time she sat on the edge, while her partner came and went, always lingering a moment to look in. It was pretty to see him making up his mind where to put the morsel, so small that it did not show in the beak. He turned his head one side and then the other, considered, decided, and at last thrust it in the selected mouth.

The resting-time of the newly made matron was short; for when those youngsters were four days old—so fast do birdlings grow—the labor of both parents was required to keep them fed. Every ten minutes of the day one of the pair came to the nest: the father invariably alighted, deliberated, fed, and then flew; while the mother administered her mouthful, and then either slipped into the nest, covering her bantlings completely, or rested upon the edge for several minutes. There was always a marked difference in the conduct of the pair.

Six days the kingbird babies were unseen from below; but on the seventh day of their life two downy gray caps were lifted above the edge of the dwelling, accompanied by two small yellow beaks, half open for what goods the gods might provide. After that event, whenever the tender mother sat on her nest, two—and later three—little heads showed plainly against her satiny white breast, as if they were resting there, making a lovely picture of motherhood.

Not for many days lasted the open-mouth baby stage in these rapidly developing youngsters. Very soon they were pert and wide awake, looking upon the green world about them with calm eyes, and opening mouths only when food was to be expected. Mouthfuls, too, were no longer of the minute order; they were large enough for the parents themselves, and of course plain to be seen. Sometimes, indeed, as in the case of a big dragonfly, the father was obliged to hold on, while the young hopeful pulled off piece after piece, until it was small enough for him to manage; occasionally, too, when the morsel was particularly hard, the little king passed it back to the giver, who stood waiting, and received it again when it had been apparently crushed or otherwise prepared, so that he could swallow it.

Midsummer was at hand. The voices of young birds were heard on every side. The young thrasher and the robin chirped in the grove; sweet bluebird and pewee baby cries came from the shrubbery; the golden-wing leaned far out of his oaken walls, and called from morning to night. Hard-working parents rushed hither and thither, snatching, digging, or dragging their prey from every imaginable hiding-place. It was woful times in the insect world, so many new hungry mouths to be filled.

All this life seemed to stir the young kings: they grew restless; they were late. Their three little heads, growing darker every day, bobbed this way and that; they changed places in the nest; they thrust out small wings;

above all and through all, they violently preened themselves. In fact, this elaborate dressing of feathers was their constant business for so long a time that I thought it no wonder the grown-up kingbird pays little attention to his dress; he does enough pluming in the nursery to last a lifetime.

On the twelfth day of their life, the young birds added their voices to the grand world-chorus in a faint, low "che-up," delivered with a kingbird accent; then, also, they began to sit up calmly, and look over the edge of the nest at what went on below, quite in the manner of their fathers. Two days later, the first little king mounted the walls of his castle, fluttered his wings, and apparently meditated the grand plunge into the world outside of home. So absorbed was he in his new emotions that he did not see the arrival of something to eat, and put in a claim for his share, as usual. I thought he was about to bid farewell to his birthplace. But I did not know him.

Not till the youngest of the family was ready to go did he step out of the nest—the three were inseparable. While I waited, expecting every moment to see him fly, there was a sudden change in the air, and very shortly a furious storm of wind and rain broke over us. Instantly every young bird subsided into the nest, out of sight; and in a few minutes their mother came, and gave them the protection of her presence.

Several days were spent by the oak-tree household in shaking out the wings, taking observations of the world, dressing the feathers, and partaking of luncheon every few minutes. Such a nestful of restlessness I never saw; the constant wonder was that they managed not to fall out. Often the three sat up side by side on the edge, white breasts shining in the sun, and heads turning every way with evident interest. The dress was now almost exactly like the parents'. No speckled bib, like the bluebird or robin infant's, defaces the snowy breast; no ugly gray coat, like the redwing baby's, obscures the beauty of the little kingbird's attire. He enters society in full dress.

But each day, now, the trio grew in size, in repose of manner, and in strength of voice; and before long they sat up hours at a time, patient, silent, and ludicrously resembling the

"Three wise men of Gotham
Who went to sea in a bowl."

In spite of their grown-up looks and manners, they did not lose their appetite; and from breakfast, at the unnatural hour of half past four in the morning, till a late supper, when so dark that I could see only the movement of feeding like a silhouette against the white clouds, all through the day, food came to the nest every two minutes or less. Think of the work of those two birds! Every mouthful brought during those fifteen and a half hours required a separate hunt. They usually flew out to a strip of low land, where the grass was thick and high. Over this they hovered with beautiful motion, and occasionally dropped an instant into the grass. The capture made, they started at once for the nest, resting scarcely a moment. There were thus between three and four hundred trips a day, and of course that number of insects were destroyed.

Even after the salt bath, which one bird took always about eleven in the morning, and the other about four in the afternoon, they did not stop to dry their plumage; but simply passed the wing feathers through the beak, paying no attention to the breast feathers, which often hung in locks, showing the dark part next the body, and so disguising the birds that I scarcely knew them when they came to the nest.

The bath was interesting. The river, so called, was in fact an arm of the Great South Bay, and of course salt. To get a bath, the bird flew directly into the water, as if after a fish; then came to the fence to shake himself. Sometimes the dip was repeated once or twice, but more often bathing ended with a single plunge.

Two weeks had passed over their heads, and the three little kings had for several days dallied with temptation on the brink before one set foot outside the nest. Even then, on the fifteenth day, he merely reached the door-step, as it were, the branch on which it rested. However, that was a great advance. He shook himself thoroughly, as if glad to have room to do so. This venturesome infant hopped about four inches from the walls of the cottage, looked upon the universe from that remote point, then hurried back to his brothers, evidently frightened at his own boldness.

On the day of this first adventure began a mysterious performance, the meaning of which I did not understand till later, when it became very familiar. It opened with a peculiar call, and its object was to rouse the young to follow. So remarkable was the effect upon them that I have no doubt a mob of kingbirds could be brought together by its means.

It began, as I said, with a call, a low, prolonged cry, sounding, as nearly as letters can express it, like "Kr-r-r-r! Kr-r-r-r!" At the same moment, both parents flew in circles around the tree, a little above the nest, now and then almost touching it, and all the time uttering the strange cry. At the first sound, the three young kings mounted the edge, wildly excited, dressing their plumage in the most frantic manner, as if their lives depended on being off in an instant. It lasted but a few moments: the parents flew away; the youngsters calmed down.

In a short time all the nestlings were accustomed to going out upon the branch, where they clustered together in a little row, and called and plumed alternately; but one after another slipped back into the dear old home, which they apparently found it very hard to leave. Often, upon coming out of the house, after the imperative demands of luncheon or dinner had dragged me for a time away from my absorbing study, not a kingbird, old or young, could be seen. The oak was deserted, the nest perfectly silent.

"They have flown!" I thought.

But no: in a few minutes small heads began show above the battlements; and in ten seconds after the three little kings were all in sight, chirping and arranging their dress with fresh vigor, after their nap.

Not one of the young family tried his wings till he was seventeen days old. The first one flew perhaps fifteen feet, to another branch of the native tree, caught at a cluster of leaves, held on a few seconds, then scrambled to a twig and stood up. The first flight accomplished! After resting some minutes, he flew back home, alighting more easily this time, and no doubt considered himself a hero. Whatever his feelings, it was evident that he could fly, and he was so pleased with his success that he tried it again and again, always keeping within ten or fifteen feet of home.

Soon his nest-fellows began to follow his example; and then it was interesting to see them, now scattered about the broad old tree, and then, in a little time, all back in the nest, as if they had never left it. After each excursion came a long rest, and every time they went out they flew with more freedom. Never were young birds so loath to leave the nursery, and never were little folk so clannish. It looked as if they had resolved to make that homestead on the top branch their headquarters for life, and, above all, never to separate.

That night, however, came the first break, and they slept in a droll little row, so close that they looked as if welded into one, and about six feet from home. For some time after they had settled themselves the mother sat by them, as if she intended to stay; but when it had grown quite dark, her mate sailed out over the tree calling; and she—well, the babies were grown up enough to be out in the world—she went with her spouse to the poplar-tree.

Progress was somewhat more rapid after this experience, and in a day or two the little kings were flying freely, by short flights, all about the grove, which came quite up to the fence. Now I saw the working of the strange migrating call above mentioned. Whenever the old birds began the cries and the circling flight, the young were thrown into a fever of excitement. One after another flew out, calling and moving in circles as long as he could keep it up. For five minutes the air was full of kingbird cries, both old and young, and then fell a sudden silence. Each young bird dropped to a perch, and the elders betook themselves to their hunting-ground as calmly as if they had not been stirring up a rout in the family.

Usually, at the end of the affair, the youngsters found themselves widely apart; for they had not yet learned to fly together, and to be apart was, above all things, repugnant to the three. They began calling; and the sound was potent to reunite them. From this side and that, by easy stages, came a little kingbird, each flight bringing them nearer each other; and before two minutes had passed they were nestled side by side, as close as ever. There they sat an hour or two and uttered their cries, and there they were hunted up and fed by the parents. There, I almost believe, they would have stayed till doomsday, but for the periodical stirring up by the mysterious call. No matter how far they wandered—and each day it was farther and farther—seven o'clock always found them moving; and all three came back to the native tree for the night, though never to the nest again.

No characteristic of the young kingbirds was more winning than their confiding and unsuspicious reception of strangers, for so soon as they began to frequent other trees than the one the paternal vigilance had made comparatively sacred to them, they were the subjects of attention. The English sparrow was first, as usual, to inquire into their right to be out of their own tree. He came near them, alighted, and began to hop still closer. Not in the least startled by his threatening manner, the nearest youngster looked at him, and began to flutter his wings, to call, and to move toward him, as if expecting to be fed. This was too much even for a sparrow; he departed.

Another curious visitor was a red-eyed vireo, who, being received in the same innocent and childlike way, also took his leave. But this bird appeared to feel insulted, and in a few minutes stole back, and took revenge in a most peculiar way; he hovered under the twig on which the three were sitting, their dumpy tails hanging down in a row, and actually twitched the feathers of those tails! Even that did not frighten the little ones; they leaned far over and stared at their assailant, but nothing more. I looked carefully to see if the vireo had a nest on that tree, so strange a thing it seemed for a bird to do. The tree in question was quite tall, with few branches, an oak grown in a close grove, and I am sure there was no vireo nest on it; so that it was an absolutely gratuitous insult.

In addition to supplying the constantly growing appetites of the family, the male kingbird did not forget to keep a sharp lookout for intruders; for, until the youngsters could take care of themselves, he was bound to protect them. One day a young robin alighted nearer to the little group than he considered altogether proper, and he started, full tilt, toward him. As he drew near, the alarmed robin uttered his baby cry, when instantly the kingbird wheeled and left; nor did he notice the stranger again, although he stayed there a long time. But when an old robin came to attend to his wants, that was a different matter; the kingbird went at once for the grown-up bird, thus proving that he spared the first one because of his babyhood.

It was not till they were three weeks old that the little kings began to fly any lower than about the level of their nest. Then one came to the fence, and the others to the top of a grape-trellis. I hoped to see some indication of looking for food, and I did; but it was all looking up and calling on the parents; not an eye was turned earthward. Now the young ones began to fly more nearly together, and one could see that a few days' more practice would enable them to fly in a compact little flock. Shortly before this they had ceased to come to the native tree at night, and by day extended their wanderings so far that sometimes they were not heard for hours. Regularly, however, as night drew near, the migrating cry sounded in the grove, and upon going out I always found them together—three

"Silver brown little birds,
Sitting close in the branches."

These interesting bantlings were twenty-four days old when it became necessary for me to leave them, as they had already left me. It was a warm morning, near the end of July, and about half an hour before I must go I went out to take my last look at them. Their calls were still loud and frequent, and I had no difficulty in tracing them to a dead twig near the top of a pine-tree, where they sat close together, as usual, with faces to the west; lacking only in length of tail of being as big as their parents, yet still calling for food, and still, to all appearances, without the smallest notion that they could ever help themselves.

Thus I left them.

III. THE BABES IN THE WOOD

The little home in the wood was well hidden. About its door were no signs of life, no chips from its building, no birds lingering near, no external indication whatever. In silence the tenants came and went; neither calls, songs, nor indiscreet tapping gave hint of the presence of woodpeckers in the neighborhood, and food was sought out of sight and hearing of the carefully secluded spot. No one would have suspected what treasures were concealed within the rough trunk of that old oak but for an accident.

Madam herself was the culprit. In carrying out an eggshell, broken at one end and of no further use, she dropped it near the foot of the tree. To her this was doubtless a disaster, but to me it was a treasure-trove, for it told her well-kept secret. The hint was taken, the home soon found in the heart of an oak, with entrance twenty feet from the ground, and close observation from a distance revealed the avian owner, a golden-winged woodpecker.

The tree selected by the shy young pair for their nursery stood in a pleasant bit of woods, left wild, on the shore of the Great South Bay, "where precious qualities of silence haunt," and the delicious breath of the sea mingled with the fragrance of pines. One must be an enthusiast to spy out the secrets of a bird's life, and this pair of golden-wings made more than common demand on the patience of the student, so silent, so wary, so wisely chosen, their sanctum. Before the door hung a friendly oak branch, heavy with leaves, that swayed and swung with every breeze. Now it hid the entrance from the east, now from the west, and with every change of the vagrant wind the observer must choose a new point of view.

Then the birds! Was ever a pair so quiet? Without a sound they came, on level path, to the nest, dropped softly to the trunk, slipped quickly in, and, after staying about one minute inside, departed as noiselessly as they came. Their color, too! One would think a bird of that size, of golden-brown mottled with black, with yellow feather-shafts and a brilliant scarlet head-band, must be conspicuous. But so perfectly did the soft colors harmonize with the rough, sun-touched bark, so misleading were the shadows of the leaves moving in the breeze, and so motionless was the bird flattened against the trunk, that one might look directly at it and not see it.

For a few days the woodpeckers were so timid that I was unable to secure a good look at them. The marked difference of manner, however, convinced me that both parents were engaged in attending upon the young family; and as they grew less vigilant and I learned to distinguish them, I discovered that it was so. The only dissimilarity in dress between the lord and lady of the golden-wing family is a small black patch descending from the beak of the male, answering very well to the mustache of bigger "lords of creation."

In coming to the nest, one of the pair flew swiftly, just touched for an instant the threshold, and disappeared within; this I found to be the head of the household. The other, the mother, as it proved, being more cautious, alighted at the door, paused, thrust her head in, withdrew it, as if undecided whether to venture in the presence of a stranger, and, after two or three such movements, darted in. Always in one minute the bird reappeared, flew at once out of the wood, at about the height of the nest, and did not come down till it reached, on one side, an old garden run to waste, or, on the other, far over the water, a cultivated field. At that tender age, the young flickers received their rations about twice in an hour.

Although the golden-wings were silent, the wood around them was quite lively from morning till night. Blackbirds and cuckoos flew over; orioles, both orchard and Baltimore, sang and foraged among the trees; song-sparrows and chippies trilled from the fence at one side: bluebird and thrasher searched the ground, and paid in music for the privilege; pewees and kingbirds made war upon insects; and from afar came the notes of redwing and meadow-lark. Others there were, casual visitors, and of course it did not escape the squawks and squabbles of the English sparrow—

"Irritant, iterant, maddening bird."

The robins, who one sometimes wishes, with Lanier's owl, "had more to think and less to say," were not so self-assertive as they usually are; in fact, they were quite subdued. They came and went freely, but they never questioned my actions, as they are sure to do where they lead society. Now and then one perched on the fence and regarded me, with flick of wing and tail that meant a good deal, but he expressed no opinion. With kingbirds on one side, pewees on the other, and the great crested fly-catcher a daily caller, this was eminently a fly-catcher grove, and the robin plainly felt that he was not responsible for its good order. Indeed, after fly-catcher households were set up, he had his hands full to maintain his right to be there at all.

Whatever went on, the woodpeckers took no part in it. Back and forth they passed, almost stealthily, caring not who ruled the grove so that their precious secret was not discovered. Neither of them stayed to watch the nest, nor did they come and go together. The birds in the neighborhood might be inquisitive—there was no one to resent it; blackbirds scrambled over the oak, robins perched on the screening branch, and no one about the silent entrance disputed their right.

In the first flush of dismay at finding themselves watched, the golden-wings, as I said, redoubled their cautiousness. They tried to keep the position of the nest secret by coming from the back, gliding around on the trunk, and stealing in at the door, or by alighting quietly high up in the body of the tree, and coming down backward—that is, tail first. But by remaining absolutely without motion or sound while they were present, I gradually won their toleration, and had my reward. The birds ceased to regard me as an enemy, and, though they always looked at me, no longer tried to keep out of sight, or to hide the object of their visits. During the first day of watching I had the good fortune to see a second empty shell brought out of the nest, and dropped a little farther off than the first had been; and I feel safe in assuming that these two were the birthdays of the babes in the wood.

Thirteen days were devoted to the study of the manners and customs of the parents before the hidden subjects of their solicitude gave any signs of life visible from below. Though visits were about half an hour apart, and flicker babies have very good appetites, they did not go hungry, for on every occasion they had a hearty meal instead of the single mouthful that many young birds receive.

This fact was guessed at on the thirteenth day, when the concealed little ones came out of the darkness up to the door, and the parents' movements in feeding could be seen; but the whole curious process was plain two days later, when a young golden-wing appeared at the opening and met his supplies half-way. The food-bearer clung to the bark beside the entrance, leaned over, turned his head on one side, and thrust his beak within the slightly opened beak of his offspring. In this position he gave eight or ten quick little jerks of his head, which doubtless represented so many mouthfuls; then, drawing back his head, he made a motion of the throat, as though swallowing, which was, presumably, raising instead, for he leaned over again and repeated the operation in the waiting mouth. This performance was gone through with as many as three or four times in succession before one flicker baby was satisfied. After the nestlings came up to the door, the parents went no more inside, as a rule, and housekeeping took care of itself.

On the fifteenth day of his life, as said above, the eldest scion of the golden-wing family made his appearance at the portal of his home. The sight and the sound of him came together, for he burst out at once with a cry. It was not very loud, but it meant something, and the practice of a day or two gave it all the strength that was desirable. In fact, it became clamorous to a degree that made further attempts at concealment useless, and no one was quicker to recognize it than the parents.

The baby cry was the utterance familiar from the grown-up birds as "wick-a! wick-a! wick-a!" From this day, when one of the elders drew near the tree, it was met at the opening by an eager little face and a begging call; but it was several days before the recluse showed interest in anything except the food supply. Meals were now nearly an hour apart, and the moment one was over the well-fed youngster in the tree fell back out of sight, probably to sleep, after the fashion of babies the world over. But all this soon came to an end. The young flicker began to linger a few minutes after he had been fed, and to thrust his beak out in a tentative way, as if wondering what the big out-of-doors was like, any way.

Matters were going on thus prosperously, when a party of English sparrows, newly fledged, came to haunt the wood in a small flock of eighteen or twenty; to meddle, in sparrow style, with everybody's business; and to profane the sweet stillness of the place with harsh squawks. The mistress of the little home in the oak, who had conducted her domestic affairs so discreetly, one day found herself the centre of a mob; for these birds early learn the power of combination. She came to her nest followed by the impertinent sparrows, who flew as close as possible, none of them more than a foot from her. They alighted as near as they could find perches, then crowded nearer, stretched up, flew over, and tried in every way, with an air of the deepest interest, to see what she *could* be doing in that hole.

When she left—which she did soon, for she was annoyed—the crowd did not go with her; they were bound to explore the mystery of that opening. They flew past it; they hovered before it; they craned their necks to peer in; they perched on a bare twig that grew over it, as many as could get footing, and leaned far over to see within. The young flicker retired before his inquisitive visitors, and was seen no more till the mother came again; and then she had to go in out of sight to find him.

As the days went on, the babe in the wood became more used to the sunlight and the bird-sounds about him. Evidently, he was of a meditative turn, for he did not scramble out, and rudely rush upon his fate; he deliberated; he studied, with the air of a philosopher; he weighed the attractions of a cool and breezy world against the comforts and delightful obscurity of home. Perhaps, also, there entered into his calculations the annoyance of a reporter meeting him on the threshold of life, tearing the veil away from his private affairs. What would one give to know the thoughts in that little brown head, on its first look at life! Whatever the reason, he plainly concluded not to take the risk that day, for he disappeared again behind a door that no reporter, however glib or plausible, could pass. Sometimes he vanished with a suddenness that was not natural. Did his heart fail him, or, perchance, his footing give way? For whether he clung to the walls, or made stepping-stones of his brothers and sisters (as do many of his betters, or at least his biggers), who can tell? Often beside this eldest-born, after the first day, appeared a second little head, spying eagerly, if a little less bravely, on the world, and as days passed he frequently contested the position of vantage with his brother, but he was always second.

Mother Nature is kind to woodpeckers. She fits them out for life before they leave the seclusion of the nursery. There is no callow, immature period in the face of the world, no "green" age for the gibes or superior airs of elders. A woodpecker out of the nest is a woodpecker in the dress and with the bearing of his fathers—dignified, serene, and grown up.

As the sweet June days advanced, the young bird in the oak-tree grew bolder. He no longer darted in when a saucy sparrow came near, and when the parent arrived with food the cries became so loud that all the world could know that here were young woodpeckers at dinner. Now, too, he began to spend much time in dressing his plumage, in preparation for the grand début. Usually, when a young bird begins to dally with the temptation to fly, so rapid is growth among birds, he may be expected out in a few hours. In this deliberate family it is different; indeed, taking flight must be a greater step for a woodpecker than for a bird from an open nest.

Three days the youngster had been debating whether it were "to be or not to be," and more and more he lingered in the doorway, sitting far enough out to show his black necklace. His was no longer the wondering gaze of infancy, to which all things are equally strange; it was a discriminating look—the head turned quickly, and passing objects drew his attention. On the third day, too, he uttered his first genuine woodpecker cry of "pe-auk!" He had not the least embarrassment before me. I think he regarded me as a part of the landscape—the eccentric development of a tree trunk, perhaps; for while he never looked at me nor put the smallest restraint upon his infant passions, let another person come into the wood, and he was at once silent and on his guard. All this time he had become more and more fascinated with the view without his door; one could fairly see the love of the world grow upon him. He picked at the bark about him; he began to get ideas about ants, and ran out a long tongue and helped himself to many a tidbit.

When the young golden-wing had passed four days in this manner, he grew impatient. The hour-long intervals between meals were not to his mind, and he began to express himself fluently. He leaned far out, and delivered the adult cry with great vigor and new pathos; he then bowed violently many times, moved his mouth as if eating, and struggled farther and still farther out, until it seemed that he could not keep within another minute. When one of the parents came he forgot his grown-up manner, and returned to the baby cry, loud and urgent, as if he were starved.

He was fed, and again left; and now he scrambled up with his feet on the edge. He was silent; he was considering an important move, a plunge into the world. He wanted to come—he longed to fly. Outside were sunshine, sweet air, trees, food—inside only darkness. The smallest coaxing would bring him out; but coaxing he was not to have. He must decide for himself; the impulse must be from within.

The next morning opened with a severe northeast gale.

"It rained, and the wind was never weary."

The birds felt the depressing influence of the day. The robins perched on the fence, wings hanging, each feather like a bare stick, and not a sound escaping the throat; and when robins are discouraged, it is dismal weather indeed. The bluebirds came about, draggled almost beyond recognition. Even the swallows sailed over silently, their merry chatter hushed.

But life must go on, whatever the weather; and fearing the young woodpecker might select this day to make his entry into the big world, his faithful watcher donned rainy-day costume, and went out to assist in the operation. The storm did not beat upon his side of the tree, and the youngster still hung out of his hole in the trunk, calling and crying, apparently without the least intention of exposing his brand-new feathers to the rain.

Very early the following morning, before the human world was astir, loud golden-wing cries, and calls, and "laughs" were heard about the wood. This abandonment of restraint proclaimed that something had happened; and so, indeed, I discovered; for in hastening to my post I found an ominous silence about the oak-tree. The young wise-head, whose struggles and temptations I had watched so closely, had chosen to go in the magical morning hours, when the world belongs entirely to birds and beasts. The home in the wood looked deserted.

I sat down in silence and waited, for I knew the young flicker could not long be still. Sure enough, I soon heard his cry, but how far off! I followed it to an oak-tree on the farther edge of the grove. I searched the tree, and there I saw him, quiet now as I approached, and plainly full of joy in his freedom and his wings.

I returned to my place, hoping that all had not gone. There must be more than one, for two had been up to the door, I was sure. I waited. Some hours later, the parents came to their home in the wood, one after the other. Back one alighted beside the door, glanced in, in a casual way, but did not put the head in, and then flew to a neighboring tree, uttering what sounded marvelously like a chuckling laugh, and in a moment left the grove. Did, then, the daughters of the house meekly fly, without preliminary study of the world from the door? Were there, perchance, no daughters? Indeed, had more than one infant reached maturity? All these questions I asked myself, but not one shall I ever be able to answer.

I waited several hours. Many birds sang and called among the trees, but no sound came from the oak-tree household, and to me the wood was deserted.

IV. HOME LIFE OF THE REDSTART

The redstart himself told me where his treasures were "hid in a leafy hollow." Not that he intended to be so confiding; on the contrary he was somewhat disconcerted when he saw what he had done, and tried his best to undo it by appearing not to have the smallest interest in that particular tree. I happened that morning to be wandering slowly along the edge of a tree-lined ravine, looking for the nest of a greatly disturbed pair of cat-birds. As I drew near an old moss-covered apple-tree, I heard a low though energetic "phit! phit!" and a chipping sparrow emerged from the tree with much haste, quickly followed by a redstart, with the unmistakable air of proprietor. The sight of me made a diversion. The pursued dropped into the grass, while the pursuer turned his attention to the bigger game, presented so unexpectedly that he had not time to bethink himself of his usual custom of not showing his gorgeous black and gold about home. He scolded me well for an instant, till his wits returned, when he disappeared like a flash. It was too late to deceive me, however, and I marked that tree as I passed, intent at the moment upon cat-birds.

On returning, I stopped on the bank to look the tree over at my leisure, and there I soon saw, two feet from the top of the tallest upright branch and tightly clinging to it, a small cradle, gently rocking in the warm breeze. No one was at home, and I sat down to wait. This movement did not meet the approval of a certain small tenant of a neighboring tree, for I was saluted by a sharp, low, incessant cry; now it came from the right side, now from the left. I turned quickly, caught a glimpse of yellow, the flit of a wing, and then—nothing. In a moment the sound began again, and thus it tantalized me till my neck became tired, and I laid my head back among the ferns, to wait till the small fire-brand calmed down a little. To my surprise and delight, the bird seemed to regard this as a surrender, for down a broad branch that sloped toward me came a most animated bundle of feathers, wings and tail wide spread, making hostile demonstrations, and scolding as fiercely as such an atom could. It had all the airs of ownership, and its colors were olive and yellow; had, then, the roguish redstart deceived me, after all? Thus pondering, I suddenly remembered that I had never seen his spouse, and that monsieur and madame do not dress alike in the bird world any more than in the human. I marked the points; I consulted the books; there could be no doubt this was the little dame herself, and her mate had been too clever to come to her aid.

The structure on the apple bough was the redstart homestead. Watch it every day I must, yet not to disturb the fiery little owners it was necessary to move further from them. I sought and found a delightful nook, the other side of the ravine. On its steep sides the native forest still flourished, and seated at the foot of a tall maple, tented in by a heavy low growth at my back, I could look across the narrow chasm through a gap in the trees, and see the redstart nest in the pasture beyond. The restless pair did not notice me behind my veil of greenery, and my glass was of the best; so I secured a good view of the small mansion and the life that went on about it, without in the least annoying the builders thereof. I found the head of the family very interesting in his role of husband and father.

Perhaps not every one knows a redstart, and his name is misleading, for he has not a red feather on his body. He is a bird of very few inches, clothed in brilliant array of orange and black and white, which always suggests the Baltimore oriole. His mate is more soberly clad in olive-brown and golden-yellow; neither of them is still for even an instant, diving and flitting about on a tree like specks of animated sunlight.

At my pleasant post of observation I spent hours of every day, stealing in soon after breakfast, quietly, so as not to arouse the suspicions of a robin who lived in the neighborhood; for unfortunate is the student whose ways are not acceptable to one of this noisy family. I found, however, when my patience gave out, that the robin will take a hint. On throwing a pebble through the branches near him, as a suggestion that his attentions were not welcome, he flew to a tree a little farther off, and resumed his offensive remarks; another pebble convinced him that the distance might be profitably increased, and thus I drove him away; at about the fourth pebble he took a final departure.

Here, then, I saw the small housekeeping go on. I always found the little dame in possession, and generally the lord and master gleaning food in redstart fashion; flitting around a branch, darting behind a leaf, over and under a twig, tail spread to keep his balance during these jerky movements, his bright oriole colors flashing as he dashed through a patch of sunlight—a beautiful object, but a perfectly silent one. When his happiness demanded expression he flew to a maple-tree, and poured out his soul in the quaint though not very musical ditty of his race. Sometimes he stood still on a branch, like a bird who has something to say; but more often he rushed around after insects on this tree, and threw in the notes between the firm snaps of his beak.

Promptly every half hour the little sprite took his way to that precious apple branch, and dropped, light as a snow-flake, on a certain twig on the nearest side of his homestead. A flash from the nest announced the departure of madame, and he popped into her place. Not to settle down to business, as she did—far from it! It is a wonder to me how even a female redstart can sit still. On taking his place, he first examined the treasures it held, leaning over the edge with a solicitude charming to see; and when he did at last cover them from sight, his black velvet cap still bobbed up and down, this way and that, as though he were taking advantage of his enforced quiet to plume himself. Precisely three minutes he allowed his modest spouse for her repast. At the expiration of that time he deserted, darted away, and began to call from the next tree, when she instantly returned.

Sometimes she was at hand, and alighted on a twig on the farther side of the nest, when he bounded off and out of sight. She carefully inspected the nest to see that all was right, then slipped in, settled herself with a gentle flutter of wings, and I knew she was safe for another half hour. It was the closest watching I ever tried, so quick were the motions, so silent the going and coming.

Now and then the redstart chose to stay longer at home. The usual time having expired, the little sitter appeared, but if her mate did not vacate, she availed herself of the additional liberty in flitting about the tree, adding a dessert to her dinner. On one occasion he let her return twice before he left, occupying her place for eight minutes—an enormous length of time for a redstart. More often he grew impatient in less than three minutes, and once he forgot himself so far as to call while in the nest.

During the sitting there came two days of steady, pouring rain and high wind. I feared the hopes of that family, as well as others all about, would perish, but the brave little mother bore the depressing season well. The eggs were never left uncovered, nor did that gay rover, her spouse, forget to take her place as usual.

On the morning of my fourth day of watching, I saw there was news; sitting was over, and though they could not be seen, it was easy to picture the featherless, wide-mouthed objects, evidently so lovely to the young parents.

Close work as it had been to observe the movements of the pair, it was much harder after that, they became at once so wary. I am sure they never regarded me in any way as a spy, for I was not in their highway; moreover, they would certainly have expressed their mind if they had. Yet they came and went entirely from the other side, and so exactly opposite the nest that often I could not see even the flit of a wing. Not until one stood on the threshold could I see it, and the most untiring vigilance was necessary. Even before this madame was cautious in her going and coming; she first dropped about two feet to a branch, paused a moment, then went to a second one, still lower, thus left the tree near the ground, and in returning she began at the lowest branch and retraced her steps to the nest.

That day the father of the new family seemed very joyous, and treated us to a great deal of singing, though it was not a singing-day, being very cold, with a steady rain. The pretty little mother took thoughtful care of her brood. For a half hour or more she worked very busily, her mate helping, and fed them well; then she deliberately sat down upon those youngsters, exactly as though they were still eggs. There she stayed as long as she thought best, and then she went to her work again.

The morning they were six days old I had the pleasure of seeing a movement in the nest. When the sun reached a certain height above the tree, it shone into that small mansion in such a way as to reveal its contents; thus I could see the redstart babies moving restlessly, evidently in haste already to come out into the world. This day the father took rather more than half the charge of the provision supply, and with considerable regularity. During four hours that the nest was closely watched, its tenants were fed at about five-minute intervals for half an hour; and then mamma promptly smothered their ambition, as above mentioned, for perhaps a quarter of an hour, when, if they did not take naps like "good little birdies," they at least were forced to keep still.

This young matron reminded me of some mothers of a larger growth, she was so fussy, so careful that her charges did not go too fast for their strength, while her spouse made it his business to see that she did not keep them tender by over-coddling. He allowed her to brood them for fifteen minutes; longer than that he would not tolerate, but came like a fiery meteor to see that she moved. She plainly understood his intention, for the instant he appeared she darted off, although he did not touch the nest. All day the weight of responsibility kept this rover at home; he might generally be seen on the lower branches of his tree, darting about in perfect silence; but once or twice I saw him actually loitering, a pleasant pastime of which I never suspected a redstart.

Six days appears to be the limit of time a redstart baby can submit to a cradle. (I know this does not agree with the books, so I explain that it was six days from the time constant sitting ceased. If the young were out of the shell before that, they were covered all the time, and not fed.) The day that stirring urchin was six days old he mounted the edge of the nest and tried his wings. When mamma came, he asked for food in the usual bird-baby way, gentle flutters of the wings; but this haste was certainly not pleasing to the little dame, and upon her departure I noticed that he had returned to the nursery.

However, his ambition was roused—the ambition of a redstart to be moving—and at seven o'clock the next morning, his seventh day, he came out with his mind made up to stay. First a shaky little yellowish head appeared above the nest; then the owner thereof clambered out upon a twig, three inches higher. One minute he rested, to glance around the new world, and quickly increased the distance to six inches, where he stood fidgeting, arranging his feathers, and evidently preparing for a tremendous flight, when his anxious parent returned. Plainly, he would have been wiser to wait another day, for all the time it was difficult for him to keep his place; every few seconds he made wild struggles, beating the air with his wings, and at last, after enjoying that elevated position in life about ten minutes, he lost his hold and fell.

I held my breath, for a fall to the ground meant a dead nestling; but he clutched at a twig two or three feet lower, and succeeded in retaining this more humble station. Madame came and fed and comforted him, and it was soon evident that he had learned a lesson, for he moderated his transports; though his head was as restless as ever, his feet were more steady; he did not fall again, and he soon scrambled freely all over the tree.

Now I was interested to see how the redstart babies were brought up, and for more than four hours I kept my eyes on that youngster. It is no small task, let me say, to keep watch of an atom an inch or two long, to whom any leaf is ample screen, to note every movement lest he slip out of sight, and to make memorandum of each morsel of food he gets. There were, also, of course, the most seductive sounds about me; never so many birds came near. Cat-birds whispered softly behind my back; a vireo cried plaintively over my head; the towhee bunting boldly perched on a low bush, and saluted me with his peculiar cry; flickers uttered their quaint "wick-up" on my right, and a veery sighed softly "we-o" on my left. Unflinchingly, however, I kept my face toward that apple-tree, and my eyes on that restless young hopeful, while I noted the conduct of the parents toward him.

This is what I learned: first, that those left in the nest were to be kept back, and not allowed out of the nursery till this one was able to care for himself; at least to help. The nest, holding probably one or two little ones, was visited, the first hour almost exactly once in twenty minutes, by madame exclusively, and the three succeeding hours at longer intervals, by her spouse. Scarcely a move was made there; plainly there were no more "come-outers" that day. The efforts of the mother were concentrated on number one, apparently, to bring him forward as fast as possible. He was, for an hour, fed every five or six minutes, the next hour only three times, and this system was kept up with perfect regularity all day.

Meanwhile, the behavior of the happy father was peculiar and somewhat puzzling, considering how solicitous he had hitherto appeared. For some time his gay coat was not to be seen, even on his favorite lower branches; and when he did come around, his mate flew at him, whether to praise or to punish him could only be guessed, for he at once disappeared before her. After two or three episodes of this sort he remained about the tree, and occasionally contributed a mite or two to the family sustenance.

The next morning, at half past seven, I resumed my seat as usual, and very soon saw I was too late. Both parents were busily flitting about the tree, but never once went near the old home; moreover, when the sun reached the magical point where he revealed the inside of the nest, lo, it was empty!

Either there had been but one other bairn, and he had got out before I did—things happen so rapidly in the redstart family—or there had been a tragedy, I could not discover which. Neither could I find a young bird on that tree, though I was sure, by the conduct of the parents, that at least one remained.

Now that no one's feelings could be hurt by the operation, I had a limb cut off the apple-tree, and the little home I had watched with so great interest brought down to me. Nothing could be daintier or more secure than that snug little structure. Placed on an upright branch, just below the point where five branchlets, a foot or more long, sprang out to shelter, and closely surrounded by seven twigs, of few inches but many leaves, it was a marvel I had been able to see it at all.

The redstarts might be lively and restless, but they were good workers. So firmly was that nest fastened to its branch, resting on one-twig and embraced by two others, like arms, that to remove it would destroy it. Strips of something like grapevine bark, with a few grass-blades and a material that looked like hornets' or other insects' nest, formed the outside, while long horse-hairs made the soft lining. Though strong and firm, it was on the sides so thin, that, as mentioned above, the movements of the young could be seen through it.

This pretty cup, around which so many hopes had centred, was of a size for a fairy's homestead—hardly two inches inside diameter, and less than two inches deep. I carried it off as a memento of a delightful June among the hills of the old Bay State.

V. WHEN NESTING IS OVER

"When the birds fly past
And the chimes ring fast
And the long spring shadows sweet shadow cast,"

comes the most attractive time of year to the bird-lover—the baby-days, when the labors and anxieties of the nest being over, proud and happy parents bring forward their tender younglings all unused to the ways of the world, and carry on their training before our eyes.

First to come upon the scene of the summer's studies was the brown thrush family. For some time the head of the household had made the grove a regular resting place in his daily round. He always entered in silence, alighted on the lowest limb of a tree, and hopped lightly, step by step, to the top, where he sang softly a few delightful and tantalizing strains. In a moment he dropped to the ground, uttering a liquid note or two as he went, and threw himself into his work of digging among the dead leaves the same suppressed vehemence he had put into his song.

Not infrequently he came into collision with a sparrow mob that claimed to own that piece of wood, and his way of dealing with them was an ever fresh satisfaction. He stood quiet, though the crouching attitude and the significant twitches of his expressive tail indicated very clearly to one who knew him that he was far from calm inside; that he was merely biding his time. His tranquil manner misled the vulgar foe; that they mistook it for cowardice was obvious. Nearer, and still nearer, they drew, surrounded him, and seemed about to fall upon him in a body, when he suddenly wheeled, and like a flash of light dashed right and left almost simultaneously, as if he had become two birds, and the impertinent enemy fairly vanished before him.

Like many another bird, however, the thrasher, although not afraid of sparrows, disliked a continual row. He had gradually ceased to come into the neighborhood, and I feared I should neither see nor (what was worse) hear him again. But one morning he presented himself with two youngsters, so brimful of joy that he quite forgot his previous caution and reserve. They perched in plain sight on the fence, and while the little ones clumsily struggled to maintain their footing, the father turned his head this side and that, jerked his tail, and uttered a low cry as touch as to say, "Can anybody beat that pair now?"

In a moment he fell to the serious work of filling their hungry mouths. Being very wide awake, the young birds readily saw where supplies came from, and then they accompanied their parent to the ground, following every step, as he dug almost without ceasing. After a tolerably solid repast of large white grubs, he slipped away from the dear coaxers, disappeared on the other side of the fence, and before they recovered from their bewilderment at finding themselves deserted, returned bearing in his beak a strawberry. The young thrush received the dainty eagerly, but finding it too big to swallow, beat it on the fence as if it were a worm. Of course it parted, and a piece fell to the ground, which the waiting parent went after, and administered as a second mouthful.

For a long time the little ones were fed on the fence, and the father was so happy that every few minutes he was forced to retire behind a neighboring tree and "make gladness musical upon the other side."

After that morning the thrasher came daily to the place, and a dessert of strawberries invariably followed the more substantial meal, but never again did he bring more than one of his family with him.One morning the brown thrush baby, who had been rapidly growing self-reliant, came alone for the first time. It was interesting to watch him, running along the tops of the pickets; searching in the hot grass till out of breath for something to eat; looking around in a surprised way, as if wondering why the food did not come; making a dash, with childlike innocence, after a strawberry he saw in the mouth of a robin, who in amazement leaped a foot in the air; and at last flying to a tree to call and listen for his sire.

That wise personage, meanwhile, had stolen silently into the grove, all dripping from his bath in the bay, and while indulging in a most elaborate dressing and pluming, had kept one eye on the infant in the grass below, apparently to see how he got on by himself. When at last the little one stood panting and discouraged, he called, a single "chirp." The relieved youngster recognized it and answered, and at once flew over to join him.

This restless young thrasher, excepting that he was perhaps somewhat lighter in color and a little less glossy of coat, looked at that moment as old as he ever would. Nothing but his ingenuous ways, and his soft baby-cry "chr-er-er" revealed his tender age. His curiosity when he found himself in an unfamiliar place or on a strange tree was amusing. He looked up and down, stretching his neck in his desire to see everything; he critically examined the tuft of leaves near him; he peered over and under a neighboring branch, and then gazed gravely around on the prospect before him. He flew with ease, and alighted with the grace of his family, on the bare trunk of a tree, the straight side of a picket, or any other unlikely place for a bird to be found. For a week he came and went and was watched and studied, but one day the strawberries were gathered in the old garden, and the beautiful brown thrush baby appeared no more.

The world was not deserted of bird voices, however.

> "Swift bright wings flitted in and out
> And happy chirpings were all about."

For days the wood had resounded with the shrill little cries of swallow babies, who alighted on the low trees on the border while their busy parents skimmed over the bay, or the marshy shore, and every few minutes brought food to their clamorous offspring. I had an excellent opportunity to make the acquaintance of this youngster —the white-bellied swallow. There were dozens of them, and the half grown trees were their chosen perches. The droll little fellows, with white fluffy breasts, no feet to speak of, and

> "Built so narrow
> Like the head of an arrow
> To cut the air,"

did not even notice me in my nook under the pines.

They could fly very well, and now and then one followed the parent far out, calling sharply his baby "cheep" and trying to get close to her in the air. Often she turned, met and fed him on the wing, and then sailed on, while the youngster lagged a little, unable to give his mind to feeding and flying at the same time. Sometimes the mother avoided a too persistent pleader by suddenly rising above him. When a little one was at rest, she usually paused before him on wing only long enough to poke a mouthful into his wide open beak; occasionally—but not often—she alighted beside him for a few moments.

Leading out into the water for the use of boatmen, was a narrow foot pier, provided on one side with a hand rail. This rail was a convenient rendezvous for all the babies belonging to the swallow flock, a sort of a community nursery. On this they rested from the fatigue of flying; here they were fed, and sometimes gently pushed off the perch afterward, as a mild hint to use their wings.

I wanted to find out whether parents and young knew each other from all the rest. Of course in this crowd it was not possible to tell, but I found a better chance in another favorite spot, an old post that rose out of the water, eight or ten feet from the shore, and so small that it was only comfortable for one, although two could stand on it. The post seldom lacked its occupant, a baby swallow with head up, looking eagerly into the flock above him. This isolated youngling I made my special study. Sometimes on the approach of a grown up bird, he lifted his wings and opened his mouth, petitioning for, and plainly expecting food. At other times he paid not the least attention to a swallow passing over him, but sat composed and silent, though watchful, apparently for the right one to come in sight.

He was often, though not invariably, fed upon his appeal; but that proves nothing, for it would require the services of a dozen parents to respond to every request of a young bird. It not infrequently happened, too, that one of the flock always flying about over the water came very near the little one on the post as if to offer him a morsel, but suddenly, when almost upon him, wheeled and left—obviously mistaken.

On no such occasion did that knowing youngster show any expectation of attention. Again there would sometimes join him on the post, a second young swallow, and, although crowded, they were quite contented together. Then I noticed as the elders swept over, that sometimes one baby begged, sometimes the other; never both at once. This seemed to indicate that the little one knows its parents, for no one familiar with the craving hunger and the constant opening of the baby beak to its natural purveyors, will doubt that when a young bird failed to ask, it was because the elder was not its parent.

An early lesson in many bird lives is that of following, or flying in a flock, for at first the babies of a brood scatter wildly, and seem not to have the smallest notion of keeping together. The small swallows in the trees near me were carefully trained in this. Often while one stood chirping vehemently, clearly thinking himself half starved, a grown-up bird flew close past him, calling in very sweet tones, and stopped in plain sight, ten or fifteen feet away. Of course the youngster followed at once. But just as he reached the side of the parent, that thoughtful tutor took another short flight, calling and coaxing as before. This little performance was repeated three or four times before the pupil received the tidbits he so urgently desired.

Other sweet baby-talk in the trees came from the wood-pewee. The pewee I had noted from the building of her beautiful lichen-covered cradle in the crotch of a wild-cherry tree. The branch, dead and leafless, afforded no screen for the brave little mother. Look when one might, in the hottest sunshine or the heaviest rain, there sat the bird quite up out of the nest, head erect and eyes eagerly watching for intruders. The pewee, for all his tender and melancholy utterances, has a fiery spirit. He hesitates not to clinch with a brother pewee, interpolates his sweetest call into the hot chases, and even when resting between encounters, spreads his tail, flutters his wings, and erects his crest in a most warlike manner. The little dame was not a whit less vigilant than her spouse. Let but a blackbird pass over and she was off in a twinkling, pursuing him, pouncing down upon him savagely, and all the time uttering her plaintive "pe-o-wee!" till her mate joined her, and made it so uncomfortable for the big foe that he departed, protesting to be sure in vigorous black-birdese, but taking good care to go. So persistent were the pewees in these efforts, that in a few days they convinced a pair of blackbirds (purple crow blackbirds) that this part of the grove was no longer a thoroughfare, and whereas they had been quite frequent visitors, they were now rarely seen.

The saucy robin who chose to insist upon his right to alight on their tree, as he had always done, was harder to convince; in fact, he never was driven away. Every day, and many times a day, arose the doleful cry of distress. I always looked over from my seat on the other side of the little open spot in the wood, and invariably saw a robin on the lower part of the wild-cherry where the trunk divided, flirting his tail, jerking his wings, and looking very wicked indeed. Down upon him came one, sometimes two pewees. He simply ran up the sloping branch toward their nest, hopped to another limb, every step bringing him nearer, the pewees darting frantically at him—and at last took flight from the other side; but not until he was quite ready. This drama was enacted with clock-like regularity, neither party seeming to tire of its repetition, till the happy day when the pewee baby could fly, and appeared across the grove, near me.

One morning I noticed the anxious parents very busy on a small oak-tree, but a clump of leaves made a perfect hiding place for the infant, and I could not see it at first. There may have been more, although I saw but one and heard but one baby cry, a prolonged but very low sound of pewee quality. While their charge lingered so near me, I was treated to another sensation by one of the pair—a pewee song. The feathered performer alighted almost directly over my head, and began at once to sing in a very sweet voice, but so low it could not be heard a dozen feet away. There was little variation in the tones, but it was rapidly delivered, with longer and shorter intervals and varying inflections, a genuine whisper-song such as most birds that I have studied delight in. It did not please madam, his mate; she listened, looked, and then rushed at the singer, and I regret to say, they fell into a "scrimmage" in the grass, quite after the vulgar manner of the sparrow.

They soon returned to their duty of feeding the baby behind the oak leaf screen. Both came very nearly at the same time; each one on arriving, administered a significant "poke" behind the leaf, then indulged in several eccentric movements in their jerky style, dashed after a fly, stood a full minute staring at me, and at last flew. This programme was scarcely varied. Inoffensive as I was, however, the birds plainly did not relish my spying upon them, and when I returned from luncheon, they had removed their infant. For a day or two, I heard on the farther side of the grove the sweet, mournful "pe-o-wee" with which this bird proclaims the passage of another insect to its fate, and then it was gone, and I saw and heard them no more.

One morning I rose at dawn and seated myself behind my blind to spy upon the doings of the early risers. On this particular morning I first heard the tender notes of "the darling of children and bards"—the bluebird baby. The cry was almost constant; it was urgent and clamorous beyond anything I ever heard from "April's bird." I even doubted the author till I saw him. The thin and worn looking mother who had him in charge worked without ceasing, while the open-mouthed infant lifted up his voice and wept in a way so petulant and persistent as to completely disguise its normally sweet bluebird quality.

Now this charming youngster, bearing heaven's color on his wings, with speckled bib and shoulder-cape, and honest, innocent eyes, is a special favorite with me; I never before saw a cry-baby in the family, and I did not lose sight of him. Three or four days passed in which the pair frequently came about, but without the father or any other young ones. Had there been an accident and were these the survivors? Was the troublesome brawler a spoiled "only child"? All questions were settled by the appearance somewhat later of three other young bluebirds who were not cry-babies. The father had evidently shaken off the trammels of domestic life, and "gone for his holiday" into the grove, where his encounters with the pewees kept up a little excitement for him.

When the pitiful looking little dame had succeeded in shaking off her ne'er-do-well, the four little ones came every day on the lawn together. Sometimes the mother came near to see how they prospered, but oftener they were alone. They cried no more; they ran about in the grass, and if one happened upon a fat morsel, the three others crowded around him and asked in pretty baby fashion for a share. Often they went to the fence, or the lower bar of the grape trellis, and there stood pertly erect, with head leaning a little forward, as though pondering some of the serious problems of bluebird life, but in fact concerning themselves only with the movements in the grass, as now and then a sudden plunge proved. Sometimes one of the group appeared alone on the ground, when no person was about (except behind the blinds), and then he talked with himself for company, a very charming monologue in the inimitable bluebird tone, with modifications suggesting that a new and wonderful song was possible to him. He was evidently too full of joy to keep still.

The English sparrow—he who had usurped the martin house in the yard—warned him off; the golden warbler, who flitted about the shrubbery all day, threatened to annihilate him, but with infantile innocence he refused to understand hostility; he stared at his assailant, and he held his ground. The little flock of four was captivating to see, and though the mother looked ragged and careless in dress, one could but honor the little creature who had made the world so delightful a gift as four beautiful new bluebirds, in whose calm eyes

"Shines the peace of all being without cloud."

Other young birds were plentiful in those warm July days. From morning till night the chipping sparrow baby, with fine streaked breast, uttered his shrill cricket-like trill. No doubt he had already found out that he would get nothing in this world without asking, so, in order that nothing escape him, his demand was constant. The first broods of English sparrows had long before united in a mob, and established themselves in the grove, and the nests were a second time full of gaping infants calling ever for more. The energies of even this unattractive bird were so severely taxed that he spared us his comments on things in general, and our affairs in particular. In the wood, young high-holes thrust their heads out of the door and called; blackbird and martin babies flew over with their parents, talking eagerly all the way; barn swallow nestlings crowded up to the window-sill to look out and be fed by passing mothers; and cautious young kingbirds, in black caps, dressed their feathers on the edge of the nest.

But days hurried on; before long, young birds were as big as their fathers and had joined the ranks of the grown-ups. There were no more babies left on tree or lawn, and holiday time was over.

VI. IN SEARCH OF THE BLUEJAY

"The grass grows up to the front door, and the forest comes down to the back; it's the end of the road, and the woods are full of bluejays."

Such was the siren song that lured me to a certain particular nook on the side of the highest mountain in Massachusetts one June. The country was as gloriously green and fresh and young as if it had just been created. From my window I looked down the valley beginning between Greylock and Ragged Mountain, and winding around other and (to me) nameless hills till lost in the distance, apparently cut square off by what looked like an unbroken chain from east to west. The heavy forests which covered the hills ended in steep grass-covered slopes, with dashing and hurrying mountain brooks between, and, save the road, scarcely a trace of man could be seen.

The birds were already there. The robin came on to the rail fence, and with rain pouring off his sleek coat, bade us "Be cheery! be cheery!" the bluebird sat silent and motionless on a fence post; the "veery's clarion" rang out all the evening from the valley below; many little birds sang and called; and

> "The gossip of swallows filled all the sky."

But the bluejays?

The bluejays, too, were there. One saucily flirted his tail at me from the top of a tree; another sly rogue flaunted his blue robes over a wall and disappeared the other side; a third shrieked in my face and slipped away behind a tree; but one and all were far too wise to reveal their domestic secrets. I knew mysteries were on foot among them, as we know little folk are in mischief by their unnatural stillness, but I knew also that not until every jay baby was out of the nest, and there was nothing to hide, should I see that cunning bird in his usual noisy, careless role.

The peculiarity of that particular corner of nature's handiwork was that any way you went you had to climb, except east, where you might roll if you chose; in fact, you could hardly do otherwise. The first day of my hunt I started west. I climbed a hill devoted to pasture, passed through the bars, and faced my mountain. It presented a compact front of spruce-trees closely interlaced at the ground, and of course impassable. But a way opened in the midst, the path of a mountain brook, deserted now and dry. I sought an alpenstock. I abandoned all impedimenta. I started up that stony path escorted on each side by a close rank of spruce. It was exceedingly steep, for the way of a brook on this mountain-side is a constant succession of falls. I scrambled over rocks; I stumbled on rolling stones; I "caught" on twigs and dead branches; I crept under fallen tree trunks; the way grew darker and more winding. How merrily had the water rushed down this path, so hard to go up! How easy for it to do so again! Nothing seemed so natural. I began to look and listen for it.

A mysterious reluctance to penetrating the heart of the mountain by this unknown and strangely hewn path stole over me. I felt like an intruder. Who could tell what the next turn might reveal? On a fallen trunk that barred my way I seated myself to rest. The silence was oppressive; not a bird called, not a squirrel chattered, not an insect hummed. The whole forest was one vast, deep, overwhelming solitude. I felt my slightest rustle an impertinence; I could not utter a sound; surely the spirit of the wood was near! A strange excitement, almost amounting to terror, possessed me. I turned and fled—that is to say, crept—down my steep and winding stair, back to the bars where I had taken leave of civilization (in the shape of one farmhouse).

Here I paused, and again the legend of bluejays allured me. From the bars, turning sharply to one side, were the tracks of cows. The strange feeling of oppression vanished. Wherever the gentle beasts had passed, I could go, sure of finding sunny openings, grassy spots, and nothing uncanny. Meekly I followed in their footsteps; the solemn grandeur of the forest had so stirred me that even the footprint of a cow was companionable.

This path led down through a pleasant fringe of beech and birch and maple trees to a beautiful brook, which was easily crossed on stones, then up the bank on the other side into an open pasture with a scattering of spruce and other trees. Now I began to look for my dear bluejays. I disturbed the peace of a robin, who scolded me roundly from the top spire of a spruce. I started out in hot haste a dainty bit of bird life—the black and yellow warbler. I listened to the delightsome song of the field-sparrow. I heard the far-off drumming of the partridge. I walked and climbed myself tired.

Then I sat down to wait. I made a nosegay of blue violets and sweetbrier leaves; I regaled myself with wintergreens in memory of my childhood; I wrote up my note-book; but never a blue feather did I see.

The next day, between showers, I tried the north, with a guide—a visiting Massachusetts ornithologist—to show me a partridge nest with the bird sitting. We followed the ups and downs of the road for a mile, passing a meadow full of bobolinks,

"Bubbling rapturously, madly,"

climbed by a grass-grown wood road a mountain-side pasture, and reached the forest. Under a dead spruce sat my lady, in a snug bed among the fallen leaves. She was wet; her lovely mottled plumage was disarranged and draggled, but her head was drawn down into her feathers in patient endurance, the mother love triumphant over everything, even fear. We stood within six feet of the shy creature; we discussed her courage in the face of the human monsters we felt ourselves to be. Not a feather fluttered, not an eyelid quivered; truly it was the perfect love that casteth out fear.

My guide went on up to the top of Greylock; I turned back to pursue my search.

Eastward was my next trip, down toward the brook that made a valley between Greylock and Ragged Mountain. My path was under the edge of the woods that fringed a mountain stream. Not the smallest of the debt we owe the bonny brook is that it wears a deep gully, whose precipitous sides are clothed with a thick growth of waving trees—beech, white and black birches, maple, and chestnut—in refreshing and delightful confusion. The stream babbled and murmured at my side as I walked slowly down, peering in every bush for nests, and at last I parted the branches like a curtain and stepped within. It was a cool green solitude, a shrine, one of nature's most enchanting nooks, sacred to dreams and birds and—woodchucks, one of which sat straight up and looked solemnly at me out of his great brown eyes.

I sat on the low-growing limb of a tree, and was rocked by the wind outside. I forgot my object. What did it matter that I should find my bluejay? Was it worth while to go on? Was anything worth while, indeed, except to dream and muse, lulled by the music of the "laughing water"? Ah! if one were a poet!

Then the birds came. A cat-bird first, with witching low song, eying me closely with that calm, dark eye of his, the while he poured it out from a shrub,

> "Like dripping water falling slow
> Round mossy rooks, in music rare;"

a vireo, repeating over and over his few notes in tireless warble; high up in the maple tree, far across the chasm, a sweet-voiced goldfinch singing his soul away outside; and lastly, a robin, who broke the charm by a peremptory demand to know my business in his private quarters. I rose to leave him in possession. In rising I disturbed another resident, a red squirrel, who ran out on a branch and delivered as vehement a piece of mind as I ever heard, stamping his little feet and jerking his bushy tail with every word, scolding all over, to the tip of his longest hair.

I left them in their green paradise. I went to my room. I sat down in my rocker to consider.

Then the winds got up. Through the "bellows pipe," as they suggestively call the head of the valley, there poured such a gale that the birds could hardly hold on to their perches. All day long it tossed the branches, tore off leaves, beat the birds, rattled the windows, and filled the blue cover to our green bowl of a valley with clouds, even half way down the sides of the mountains themselves. And at last they began to weep, and I spent my twilight by an open window, wrapped in a shawl, listening to the

"Unrivaled one, the hermit-thrush,
Solitary, singing in the west,"

and looking out upon the hills, where I still hoped to find my bluejay.

VII. IN THE WOOD LOT

"There's bluejays a-plenty up in the wood lot," said the farmer's boy, hearing me lament my unsuccessful search for that wily bird. "There's one pair makes an awful fuss every time I passes."

I immediately offered to accompany the youth on his next trip up the mountain, where he was engaged in dragging down to our level, sunshine and summer breezes, winter winds and pure mountain air, in the shape of the bodies of trees, whose noble heads were laid low by the axes last winter. One hundred and fifty cords of beauty, the slow work of unnumbered years, brought down to "what base uses"! the most beautiful of nature's productions degraded to the lowest service—to fry our bacon and bake our pies!

The farmer did not look upon it exactly in that way; he called it "cord-wood," and his oxen dragged it down day by day. Point of view makes such a difference!

The road that wound down through the valley, skirting its hills, bridging its brooks, and connecting the lonely homestead with the rest of the human world, had on one side a beautiful border of all sorts of greeneries, just as Nature, with her inimitable touch, had placed them. It was a home and a cover for small birds; it was a shade on a warm day; it was a delight to the eye at all times. Yet in the farmer's eye it was "shiftless" (the New Englander's bogy).

The other side of the road he had "improved;" it gloried in what looked at a little distance like a single-file procession of glaring new posts, which on approaching were found to be the supports of one of man's neighborly devices—barbed wire. Rejoicing in this work of his hands on the left, he longed to turn his murderous weapons against the right side. He was labored with; he bided his time; but I knew in my heart that whoever went there next summer would find that picturesque road bristling with barbed wire on both sides. It will be as ugly as man can make it, but it will be "tidy" (New England's shibboleth), for no sweet green thing will grow up beside it. Nature doesn't take kindly to barbed wire.

The old stone wall at that time was an irresistible invitation to the riotous luxuriance of vines. Elder-bushes, with their fine cream-colored blossoms, hung lovingly over it; blackberry bushes, lovely from their snowy flowering to their rich autumn foliage, flourished beside it; and a thousand and one exquisite, and to me nameless, green things hung upon it, and leaned against it, and nearly covered it up. And what a garden of delight nestled in each protected corner of an old-fashioned zig-zag fence! Yet all these are under the ban—"shiftless."

Thanks be to the gods who sowed this country so full of stones and trees, that the army of farmers who have worried the land haven't succeeded in turning it into the abomination of desolation they admire!

And now, having relieved my mind, I'll go on with the bluejay hunt.

The next morning it was, for a rarity, fine. I started up the wood road ahead of my guide, so that I might take my climb as easily as such a thing can be taken. Passing through the bare pasture, I entered the outlying clumps of spruce which form the advance-guard of the forests on Greylock, and here my leader overtook me, urging his fiery steeds, with their empty sled. Now horned beasts have had a certain terror for me ever since an exciting experience with them in my childhood. I stood respectfully on one side, prepared to fly should the "critters" (local) show malicious intent. On they came, looking at me sharply with wicked eyes. I made ready for a rush, when, lo! they turned from me, and dashed madly into a spruce-tree, nearly upsetting themselves, and threatening to run away. We were all afraid of each other.

The mortified driver apologized for their behavior on the ground that "they ain't much used to seeing a lady up in the wood lot." I generously forgave them, and then meekly followed in their footsteps, up, up, up toward the clouds, till we reached the bluejay neighborhood. Here we parted. My escort passed on still higher, and I seated myself to see at last my bluejays.

Dead silence around me. Not a leaf stirred; not a bird peeped. I began to make a noise myself—calls and imitations (feeble) of bird-notes to arouse their curiosity;

a bluejay is a born investigator. No sign of heaven's color appeared except in the patches of sky between the leaves.

Other wood dwellers came; a rose-breasted grosbeak, with lovely rosy shield, with much posturing and many sharp "clicks," essayed to find out what manner of irreverent intruder this might be. Later his modest gray-clad spouse joined him. They circled around to view the wonder on all sides. They exchanged dubious-sounding opinions. They were as little "used to seeing a lady" as the oxen. They slipped away, and in a moment I heard his rich song from afar.

No one else paid the slightest attention to my coaxing, and I returned by easy stages to the spruces, where I had the misfortune to arouse the suspicion of a robin. Do you know what it is to be under robin surveillance? Let but one redbreast take it into his obstinate little head that you are a suspicious character, and he mounts the nearest tree—the very top twig, in plain sight—and begins his loud "Peep! peep! tut, tut, tut! Peep! peep! tut, tut, tut!"

This is his tocsin of war, and soon his allies appear, and then

"From the north, from the east,
from the south and the west,
Woodland, wheat field, corn field, clover,
Over and over, and over and over,
Five o'clock, ten o'clock, twelve, or seven,
Nothing but robin-calls heard under heaven."

No matter what you do or what you don't do. One will perch on each side of you, and join the maddening chorus, driving every bird in the neighborhood either to join in the hue and cry (as do some of the sparrows), or to hide himself from the monster that has been discovered.

I tried to tire them out by sitting absolutely motionless; but three, who evidently had business in the vicinity, for each held a mouthful of worms, guarded me to right and left and in front, and never ceased their offensive remarks long enough to stuff those worms into the mouths waiting for them.

I was not able to convince them that I had no designs on robin households, and I had to own myself defeated again. Then and there I abandoned the search for the bluejay.

VIII. THE BLUEJAY BABY

My time of triumph came, however, a little later. Birds may securely hide their nests, but they cannot always silence their nestlings. So soon as little folk find their voices, whether their dress be feathers, or furs, or French cambric, they are sure to make themselves heard and seen.

One morning, two or three weeks after I had given up the bluejay search, and consoled myself with looking after baby cat-birds and thrushes, I started out as usual for a walk. I turned naturally into a favorite path beside a brook that danced down the mountain below the house. It was near the bottom of a deep gully, where I had come to grief in my search for a veery baby.

As I passed slowly up, looking well to my steps, and listening for birds, I heard a note that aroused me at once—the squawk of a bluejay. It came from the higher ground, and I looked about for a pathway up the steep bank on my right. At the most promising point I could select I started my climb. Unfortunately that very spot had been already chosen by a small rill, a mere trickle of water, to come down. It was not big enough to make itself a channel and keep to it, but it sprawled all over the land.

Now it lingered in the cows' footprints and made a little round pool of each; then it loitered on a level bit of ground, and soaked it full; when it reached a comfortable bed between the roots of trees, it almost decided to stay and be a pond, and it dallied so long before it found a tiny opening and straggled out, that if it did not result in a pond, it did accomplish a treacherous quagmire. In fact that undecided, feeble-minded streamlet totally "demoralized" the whole hillside, and with its vagaries I had to contend at every step of my way.

I reached the top, but I left deep footprints to be turned into pools of a new pattern, and as trophy I carried away some of the soil on my dress. Of my shoes I will not speak; shall we not have souls above shoe-leather?

As soon as I recovered breath after my hasty scramble to dry ground, I started toward a thick-growing belt of spruce trees which came down from the mountain and ended in a point—one tree in advance, like the leader of an army. Here I found the bird I was seeking, a much disturbed bluejay, who met me at the door—so to speak—with a defiant squawk, a warning to come no nearer.

"Ah ha!" said I, exultingly, "are your little folk in there? Then I shall see them."

I slowly advanced; she disputed my passage at every step, but nothing was to be seen till her anxiety got the better of her discretion and she herself gave me the precious secret; she suddenly slipped through the trees to the other side, and became perfectly silent.

I could not follow her path through the tangle of trees, but I could go around, and I did. On a dead spruce wedged in among the living ones I saw the object of her solicitude; a lovely sight it was! Two young bluejays huddled close together on a twig. They were "humped up," with heads drawn down into their shoulders, and breast feathers fluffed out like snowy-white floss silk, completely covering their feet and the perch. No wonder that poor little mother was anxious, for a more beautiful pair I never saw, and to see them was to long to take them in one's hands.

Silent and patient little fellows they appeared, looking at me with innocent eyes, but showing no fear. They were a good deal more concerned about something to eat, and when their mother came they reminded her by a low peep that they were still there. She gave them nothing; she was too anxious to get them out of my sight, and she disappeared behind a thick branch.

In a moment I heard the cry of a bird I could not see. So also did the twins on the tree, and to them it meant somebody being fed; they lifted their little wings, spread out like fans their short beautiful tails, and by help of both, half hopped, half flew through the branches to the other side.

I followed, by the roundabout way again, and then I saw another one. Three bonny bairns in blue were on that dead spruce tree; two close together as before, and the third—who seemed more lively—sitting alone. He lifted his crest a little, turned his head and looked squarely at me, but seeing nothing to alarm him—wise little jay!—did not move. Then again mamma came forward, and remonstrated and protested, but only by her one argument, a squawk.

I quietly sat down and tried to make myself as much a part of the bank as possible, for I wanted the distracted dame in blue to go on with her household duties, and feed those babies. After a while she did calm down a little, though she kept one distrustful eye on me, and now and then came near and delivered a squawk at me, as if to assure me that she saw through my manœuvres, and despised them.

But I cared not at that moment for her opinion of me; she did not move my sympathies as do many birds, for she appeared insulted and angry, not in the least afraid. I wanted to see her feed, and at last I did—*almost*; she was to the last too sharp for me.

She came with a mouthful of food. Each one of the three rose on his sturdy little legs, fluttered his wings, opened his beak and cried. It was a sort of whispered squawk, which shows that the bluejay is a wary bird even in the cradle. When they were all roused and eager, the mother used that morsel as a bait to coax them through the tree again. She did not give it to either of her petitioners, but she moved slowly from branch to branch, holding it before them, and as one bird they followed, led by their appetite, like bigger folk—

"Three souls with but a single thought,
Three hearts that beat as one!"

and as I had no desire to see them die of starvation, and leave the world so much poorer in beauty, I came away and left them to their repast.

That was not the end of the bluejay episode. A few days later a young bird, perhaps one of this very trio, set out by himself in search of adventures. Into the wide-open door of the barn he flew, probably to see for what the swallows were flying out and in. Alas for that curious young bird! He was noticed by the farmer's boy, chased into a corner, still out of breath from his first flight, then caught, thrust into an old canary cage, brought to the house, and given to the bird-student.

Poor little creature! he was dumb with fright, though he was not motionless. He beat himself against the wires and thrust his beak through the openings, in vain efforts to escape. We looked at him with great interest, but we had not the heart to keep him very long. In a few minutes he was taken out of the cage in a hand (which he tried to bite), carried to the door and set free.

Away like a flash went the little boy blue and alighted in a tree beside the house. For a few moments he panted for breath, and then he opened his mouth to tell the news to whom it might concern. In rapid succession he uttered half a dozen jay-baby squawks, rested a moment, then repeated them, hopping about the tree in great excitement.

In less than thirty seconds his cries were answered. A bluejay appeared on the barn; another was seen in a spruce close by; three came to a tall tree across the road; and from near and far we heard the calls of friends trooping to the rescue.

Meanwhile the birds of the neighborhood, where the squawk of a jay was seldom heard, began to take an interest in this unusual gathering. Two cedar birds, with the policy of peace which their Quaker garb suggests, betook themselves to a safe distance, a cat-bird went to the tree to interview the clamorous stranger, a vireo made its appearance on the branches, and followed the big baby in blue from perch to perch, looking at him with great curiosity, while a veery uttered his plaintive cry from the fence below.

All this attention was too much for a bluejay, who always wants plenty of elbow room in this wide world. He flew off towards the woods, where, after a proper interval to see that no more babies were in trouble, he was followed by his grown-up relatives from every quarter. But I think they had a convention to talk it over, up in the woods, for squawks and cries of many kinds came from that direction for a long time.

IN THE BLACK RIVER COUNTRY

Where shall we keep the holiday?

Up and away! where haughty woods
Front the liberated floods:
We will climb the broad-backed hills,
Hear the uproar of their joy;
We will mark the leaps and gleams
Of the new-delivered streams,
And the murmuring river of sap
Mount in the pipes of the trees.

And the colors of joy in the bird
And the love in his carol heard.
Frog and lizard in holiday coats,
And turtle brave in his golden spots.

IX. THAT WITCHING SONG

A year or two before setting up my tent in the Black River Country, began my acquaintance with the author of the witching song.

The time was evening; the place, the veranda of a friend's summer cottage at Lake George. The vireo and redstart had ceased their songs; the cat-bird had flirted "good-night" from the fence; even the robin, last of all to go to bed, had uttered his final peep and vanished from sight and hearing; the sun had gone down behind the mountains across the lake, and I was listening for the whippoorwill who lived at the edge of the wood to take up the burden of song and carry it into the night.

Suddenly there burst upon the silence a song that startled me. It was loud and distinct as if very near, yet it had the spirit and the echoes of the woods in it; a wild, rare, thrilling strain, the woods themselves made vocal. Such it seemed to me. I was strangely moved, and filled from that moment with an undying determination to trace that witching song to the bird that could utter it.

"I'm going to seek my singer," was the message I flung back next morning, as, opera-glass in hand, I started down the orchard towards the woods. I followed the path under the apple-trees, passed the daisy field, white from fence to fence with beauty—despair of the farmer, but delight of the cottagers—hurried across the pasture beyond, skirting the little knoll on which the cow happened this morning to be feeding, crossed the brook on a plank, and reached my daily walk.

This was a broad path that ran for half a mile on the edge of the lake. Behind it, penetrated every now and then by a foot-path, was the bit of old woods that the clearers of this land had the grace to leave, to charm the eye and refresh the soul (though probably not for that reason). Before it stretched the clear, sparkling waters of Lake George, and on the other side rose abruptly one of the beautiful mountains that fringe that exquisite piece of water.

Usually I passed half the morning here, seated on one of the rocks that cropped out everywhere, filling my memory with pictures to take home with me. But to-day I could not stay. I entered one of the paths, passed into the grand, silent woods, found a comfortable seat on a bed of pine needles, with the trunk of a tall maple tree for a back, and prepared to wait. I would test Thoreau's assertion that if one will sit long enough in some attractive spot in the woods, sooner or later every inhabitant of it will pass before him. I had confidence in Thoreau's woodcraft, for has not Emerson said:

"What others did at distance hear,
And guessed within the thicket's gloom,
Was shown to this philosopher,
And at his bidding seemed to come"?

and I resolved to sit there till I should see my bird. I was confident I should know him: a wild, fearless eye, I was sure, a noble bearing, a dweller on the tree-tops.

Alas! I forgot one phrase in Thoreau's statement: "sooner or *later*." No doubt the Concord hermit was a true prophet; but how many of the inhabitants are "later"—too late, indeed, for a mortal who, unlike our New England philosopher, has such weak human needs as food and rest, and whose back will be tired in spite of her enthusiasm, if she sits a few hours on a rock, with a tree for a back.

Many of the sweet and shy residents of that lovely bit of wildness showed themselves while I waited. A flicker, whose open door was in sight, and who was plainly engaged in setting her house in order, entertained me for a long time. Silently she stole in, I did not see how. Her first appearance to me was on the trunk, the opposite side from her nest, whence she slid, or so it looked, in a series of jerks to her door, paused a few minutes on the step to look sharply at me, and then disappeared, head first, within. Quick as a jack-in-the-box, her head popped out again to see if the spy had moved while she had been out of sight, and finding all serene, she threw herself with true feminine energy into her work. The beak-loads she brought to the door and flung out seemed so insufficient that I longed to lend her a broom; but I found she had a better helper than that, a partner.

When she tired, or thought she had earned a rest, she came out, and flying to the limb above the nest, began softly calling. Never was the ventriloquial quality more plainly exhibited. I heard that low "ka! ka! ka! ka! ka!" long repeated, and I looked with interest in every direction to see the bird appear. For a long time I did not suspect the sly dame so quietly resting on the branch, and when I did it was only by the closest inspection that I discovered the slight jerk of the tail, the almost imperceptible movement of the beak, that betrayed her.

Another bird as well as I heard that call, and he responded. He was exactly like her, with the addition of a pair of black "mustachios," and it may be she told him that the strange object under the maple had not moved for half an hour, and was undoubtedly some new device of man's, made of wood perhaps, for he did not hesitate on the door-step, but plunged in at once, and devoted himself to the business in hand, clearing out, while she vanished.

But though I watched this domestic scene with pleasure, and saw and noted every feather that appeared about me, the tree-tops had my closest attention, for there I was certain I should find my rare singer. Hours passed, the shadows grew long, and sadly and slowly I took my way homewards, wishing I had a charm against fatigue, mosquitoes, and other terrors of the night, and could stay out till he came.

All through the month of June I haunted that wood, seeking the unknown. Every evening I heard him, but no sight came to gladden my eyes. I grew almost to believe it merely "a wandering voice," and I went home with my longing unsatisfied.

When next the month of roses came around, I betook myself to a spur of the Hoosac Mountains to see my birds. The evening of my arrival, as the twilight gathered, rose the call of my witching voice.

"What bird is that?" I demanded, with the usual result; no one knew. (A chapter might be written on the ignorance of country people of their own birds and plants. A chapter, did I say? A book, a dozen books, the country is full of material.)

"I shall find that bird," I said, "if I stay a year." In the morning I set out. The song had come from the belt of trees that hang lovingly over a little stream on its merry way down the mountain, and thither I turned my steps. Now, my hostess had a drove of twenty cows, wild, head-tossing creatures—"Holsteins" they were—and having half a dozen pastures, they were changed about from day to day. Driving them every morning was almost as exciting as the stampede of a drove of horses, and it seemed as if they could never reconcile themselves to the idiosyncrasies of the American woman. The pasture where they were shut for the day was as sacred from my foot as if it were filled with mad dogs. My mere appearance near the fence was a signal for a headlong race to the spot to see what on earth I was doing now.

I went into the field, looking cautiously about, and satisfying myself that the too curious foreigners were not within sight, found a comfortable seat on a bank overlooking the whole beautiful view of the brook and its waving green borders, and commanding the approach to my side of the field.

This time again my mysterious singer proved to be among the "later" ones, and after spending an hour or two there, I rose to go back, when in passing a thick-growing evergreen tree, I saw that I had created a panic. There was a flutter of wings, there were cries, and on the tree, in plain sight, the towhee bunting and his brown-clad spouse. Of course there must be some reason for this reckless display; I sought the cause, and found a nest, a mere depression in the ground, and one sorry-looking youngster, the sole survivor of the perils of the situation. Over that one nestling they were as concerned as the proverbial hen with one chicken, and they flitted about in distress while I looked at their half-fledged bantling, and hoped it was a singer to ring the delightful silver-toned tremolo that had charmed me that morning.

That evening, listening on the piazza to the usual twilight chorus, the wood-thrush far-off, the towhee from the pasture, the robins all around, I heard suddenly the "quee-o" of a bird I knew, so near that I started, and my eyes fell directly upon him, standing on the lowest limb of a dead tree, not ten feet from me.

He was so near I did not need my glass, nor indeed did I dare move a finger, lest he take flight. Several times he uttered his soft call, and then, while my eyes were fastened upon him, he began quivering with excitement, his wings lifted a little, and in a clear though low tone he uttered the long-sought song. I held my breath, and he repeated it, each time lower than before. Even at that distance it sounded far off, and doubtless many times in the woods, when I looked for it afar, it may have been over my head.

A long time; that beautiful bird sat and sang his witching evening hymn, while I listened spellbound.

It was the tawny thrush—the veery.

X. THE VEERY MOTHER

My next interview with the veery family took place the following June, at the foot of Mount Greylock, in Massachusetts. I had just returned from a walk down the meadow, put on wrapper and slippers, and established myself by the window to write some letters. Pen, ink, paper, and all the accessories were spread out before me. I dipped my pen in the ink and wrote "My Dear," when a sound fell upon my ears: it was the cry of a young bird! it was new to me! it had a veery ring!

Away went my good resolutions, and my pen with them; papers flew to right and left; hither and thither scattered the letters I had meant to answer. I snatched my glass, seized my hat as I passed, and was outdoors. In the open air the call sounded louder, and plainly came from the borders of the brook that with its fringe of trees divides the yard from the pasture beyond. It was a two-syllabled utterance like "quee wee," but it had the intermitted or tremolo sound that distinguishes the song of the tawny thrush from others. I could locate the bird almost to a twig, but nobody cared if I could. It was on the other side of the brook and the deep gully through which it ran, and they who had that youngster in charge could laugh at me.

But I knew the way up the brookside. I went down the road to the bars, crossed the water on stepping stones, and in a few minutes entered a cow-path that wandered up beside the stream. All was quiet; the young thrush no doubt had been hushed. They were waiting for me to pass by, as I often did, for that was a common walk of mine. On this log I sat one day to watch a woodchuck; a little further on was the rock from which I had peeped into a robin's nest, where one egg had been alone a week, and I never saw a robin near it.

At length I reached the path that ran up the bank where I usually turned and went to the pasture, for beyond this the cow-path descended, and looked damp and wild, as if it might once have been the way of the cows, but now was abandoned. Still all was quiet, and I thought of my letters unanswered, of my slippers, and—and I turned to go back.

Just at that moment that unlucky young thrush opened his mouth for a cry; the birds had been too sure. I forgot my letters again, and looked at the path beyond. I thought I could see a dry way, so I took a step or two forward. This was too much! this I had never before done, and I believe those birds were well used to my habits, for the moment I passed my usual bounds a cry rang out, loud, and a bird flew past my head. She alighted near me. It was a tawny thrush; and when one of those shy birds, who fly if I turn my head behind the blinds, gets bold, there's a good reason for it. I thanked madam for giving me my cue; I knew now it was her baby, and I walked slowly on.

I had to go slowly, for the placing of each foot required study. It is surprising what a quantity of water will stand on the steep sides of a mountain. Some parts of this one were like a marsh, or a saturated sponge, and everywhere a cow had stepped was a small pool. As I proceeded the thrush grew more and more uneasy. She came so near me that I saw she had a gauzy-winged fly in her mouth, another proof that she had young ones near. She called, without opening her beak, her usual "quee."

Finding a dry spot, and the cry of the baby having ceased, I sat down to consider and to wait. Then the bird seemed suddenly to remember how compromising her mouthful was, and she planted herself on a branch before my eyes, deliberately ate that fly and wiped her beak, as who should say, "You thought I was carrying that morsel to somebody, but you see I have eaten it myself; there's nothing up that path." But much as I respected the dear mother, I did not believe her eloquent demonstration. I selected another point where I could stop a minute, and picked my way to it. Then all my poor little bird's philosophy deserted her; she came close to me, she uttered the greatest variety of cries; she almost begged me to believe that she was the only living creature up that gully. And so much did she move me, so intolerably brutal did she make me feel, that for the second time I was very near to turning back.

But the cry began again. How could I miss so good a chance to see that tawny youngster, when I knew

I should not lay finger on it? I hardened my heart, and struggled a few feet further.

Then some of the neighbors came to see what was the trouble, and if they could do anything about it. A black-and-white creeper rose from a low bush with a surprised "chit-it-it-it," alighted on a tree and ran glibly up the upright branch as though it were a ladder. But a glance at the "cause of all this woe" was more than his courage could endure; one cry escaped him, and then a streak of black and white passed over the road out of sight.

Next came a redstart, himself the head of a family, for he too had his beak full of provisions. He was not in the least dismayed; he perched on a twig and looked over at me with interest, as if trying to see what the veery found so terrifying, and then continued on his way home. A snow-bird was the last visitor, and he came nearer and nearer, not at all frightened, merely curious, but madam evidently distrusted him, for she flew at him, intimating in a way that he plainly understood that "his room was better than his company."

Still I floundered on, and now the disturbed mother added a new cry, like the bleating of a lamb. I never should have suspected a bird of making that sound; it was a perfect "ba-ha-ha." Yet on listening closely, I saw that it was the very tremolo that gives the song of the male its peculiar thrill. Her "ba-ha-ha," pitched to his tone, and with his intervals, would be a perfect reproduction of it. No doubt she could sing, and perhaps she does—who knows?

Now the mother threw in occasionally a louder sort of call-note like "pee-ro," which was quickly followed by the appearance of another thrush, her mate, I presume. He called, too, the usual "quee-o," but he kept himself well out of sight; no reckless mother-love made him lose his reason. Still, steadily though slowly, and with many pauses to study out the next step, I progressed. The cry, often suppressed for minutes at a time, was perceptibly nearer. The bank was rougher than ever, but with one scramble I was sure I could reach my prize. I started carefully, when a cry rang out sudden and sharp and close at hand. At that instant the stone I had put faith in failed me basely and rolled: one foot *went in,* a dead twig caught my hair, part of my dress remained with the sharp end of a broken branch, I came to one knee (but not in a devotional spirit); I struck the ground with one hand and a brier-bush with the other, but I did not drop my glass, and I reached my goal in a fashion.

I paused to recover my breath and give that youngster, who I was persuaded was laughing at me all the time, a chance to lift up his voice again. But he had subsided, while the mother was earnest as ever. Perhaps I was too near, or had scared him out of his wits by my sensational entry. While I was patiently studying every twig on the tree from which the last cry had come, the slight flutter of a leaf caught my eye, and there stood the long-sought infant himself.

He was a few feet below me. I could have laid my hands upon him, but he did not appear to see me, and stood like a statue while I studied his points. Mamma, too, was suddenly quiet; either she saw at last that my intentions were friendly, or she thought the supreme moment had come, and was paralyzed. I had no leisure to look after her; I wanted to make acquaintance with her bairn, and I did. He was the exact image of his parents; I should have known him anywhere, the same soft, tawny back, and light under-parts, but no tail to be seen, and only a dumpy pair of wings, which would not bear him very far. The feathers of his side looked rough, and not fully out, but his head was lovely and his eye was the wild free eye of a veery. I saw the youngster utter his cry. I saw him fly four or five feet, and then I climbed the bank, hopeless of returning the way I had come, pushed my way between detaining spruces, and emerged once more on dry ground. I had been two hours on the trail.

I slipped into the house the back way, and hastened to my room, where I counted the cost: slippers ruined, dress torn, hand scratched, toilet a general wreck. But I had seen the tawny-thrush baby, and I was happy. And it's no common thing to do, either. Does not Emerson count it among Thoreau's remarkable feats that

> "All her shows did Nature yield
> To please and win this pilgrim wise;
> He found the tawny thrush's brood,
> And the shy hawk did wait for him"?

XI. THE TAWNY THRUSH'S BROOD

"He found the tawny thrush's brood," writes Emerson, in enumerating the special gifts of the nature-lover whose praise he celebrates. Whether the reference were to Thoreau or to another "forest-seer," it was certainly to a fortunate and happy man, whom I have always envied till I learned to find the shy brood myself.

I shall never forget the exciting and blissful moment when I discovered my first tawny-thrush nest. It was the crowning event of a long search.

It was not until the fourth year that I had looked for him, that I came really to know the bird, to see his family, and last of all his nest. My summer abiding-place in the Black River country was very near a bit of woods where veeries were plentiful, and I saw them at all hours, and under nearly all conditions.

My favorite seat was at the foot of a low-growing tree in the edge of the woods, where the branches hung over and almost hid me. From under my green screen I could look out into a field golden with buttercups, with scattering elms and maples, while behind me was the forest, the chosen haunt of this bird. Here, unseen, I listened to his song—

"O matchless melody! O perfect art!
O lovely, lofty voice unfaltering!"

till my soul was filled with rapture, and a longing to know him in his home relations took such possession of me that the world seemed to hold but one object of desire, a veery's nest.

Yet though the woods were full of them, so wary and so wise were the little builders that not a nest could I find. I studied the descriptions in the books; I examined the nests in a collection at hand. The books declared, and the specimens confirmed the statement, that the cradle of the tawny thrush would be found amid certain surroundings. Many such places existed in the woods, and I never passed one without seeking a nest; but always unsuccessfully, till, as June days were rapidly passing, I came to have a feeling something akin to despair when I heard the veery notes.

One day—it was Sunday afternoon—I was still grieving over the lost, or rather the unfound nest, and my friend was sitting composedly on the veranda writing letters, when restlessness seized me, and I resolved to take a quiet walk. I sauntered slowly down the road, towards the woods, of course; all roads in that charming place led to the woods.

I had nearly reached the "Sunset Corner," where I had a half-formed intention of resting and then turning back, when my eyes fell upon—but hold! I will not describe it, lest I enlighten one more collector, and aid in the robbery, perhaps the death, of one more bird-mother. Suffice it to say what I saw resembled, though not perfectly, the surroundings of a veery's nest as described in the books.

Of course there could be no nest there, I thought, yet the ruling passion asserted itself at once. It would at least do no harm to look. I left the path, walked carelessly up to the spot, and looked at it. It seemed empty of life; but as I gazed, there gradually took form a head, a pair of anxious eyes fixed upon mine, a beak pointed upward, and there was my nest! almost at my feet.

Joy and surprise contended within me. I thought not of the mother's anxiety; I stood and stared, absolutely paralyzed with delight.

But not for long. I remembered my friend who had not found the tawny thrush's nest, and with whom I must share my happiness without delay, and carefully marking the locality, not to lose what I had so accidentally found, and might so easily lose, I moved quietly away till I reached the road. Then I hurried to an opening in the trees from which the house could be seen. Here I stopped; the letter-writer looked up. I waved my green bough in triumph above my head, and with the other hand I beckoned.

"A veery's nest!" she thought at once. Away went paper and pen, and in a moment she joined me. Together we stood beside the beautiful sitting thrush, so brave, though no doubt suffering from deadly terror. Then we slowly walked away, rejoicing. It was so near the house! so easy to watch! the bird not at all afraid! All the way home we congratulated ourselves.

The next morning our first thought was of the veery's nest, and on starting out for the day we turned in that direction. Alas! the old story! The nest was overturned and thrown out of place, the leaves were trampled; there had evidently been a struggle of some kind. No birds, no eggs, not a bit of broken shell—nothing was left, except one dark brown spotted feather from a large bird, whether hawk or owl I shall never know, for neglecting to take it at the moment, it was gone when I thought of it as a witness.

Again the old longing for a nest assailed me; but I was not without hope, for I had my hint. I had found out what sort of places the veeries in this neighborhood liked. After that I never went into the woods, on whatever errand bent, but I kept my eyes open for the chosen situation. I examined dozens of promising spots, and I found nests that had been used, which proved that I was on the right track, and kept up my courage.

It was several days before another tawny-thrush cradle in use gladdened our eyes, and this was in a wild part of the woods where we seldom went. We were drawn there by the song of a tiny warbler, whose nest my friend desired to find, since it was rare; and in passing a thicket of maple saplings three feet high, she discovered a nest. She quickly parted the leaves and looked in; three young birds opened their mouths for food. "Veeries!" she exclaimed, in surprise. "What a strange place!"

This little home rested on a bare dead stick that had fallen and lodged in a living branch, and the dead leaves used by veeries in their building made it conspicuous, when the eyes happened to fall upon it; but it was so well concealed by living branches that one might pass fifty times and not see it. I describe this location, for it was very unusual.

We looked at the birdlings; we walked on till we came to the place where we turned from the path to see the warbler's little domicile. My friend passed along. I lingered a moment, for it was a lovely spot, attractive to birds as to bird-lovers, and high up in the air on the upturned roots of a fallen tree

> "an elder or two
> Foamed over with blossoms white as spray."

While I stood there admiring the brave little bush that kept on living and blooming, though lifted into an unnatural position by the tree at whose feet it had grown, some mysterious drawing made me look closely at a spot beside the road which we had passed many times without special notice. There I found our third veery nest, the mother bird sitting.

Henceforth, every morning we went up the veery road, and before each little nursery we sat us down to watch and study. It was necessary to be very quiet, the birds in the saplings were so nervous; but keeping still in the woods in summer is not the easy performance it is elsewhere, though great are the inducements. From one side comes the chirp of the winter wren, from the other, low, excited calls of veeries, and nothing but absolute quiet seems necessary to capture some of the charming secrets of their lives. Meanwhile a dancing and singing host collects around one's head. I call up my philosophy; I resolve not to care, though I shall be devoured. My philosophy stands the strain; I do not *care*; but my nerves basely fail me, and after a few moments, and a dozen stings here and there, I spring involuntarily to my feet, wildly flourish my wisp of leaves, and of course put to instant flight the actors in the drama before me.

The pair of veeries in the maple bushes were never reconciled to our visits. They called and cried in all the varied inflections of their sweet voices, and they moved uneasily about on the low branches with mouths full of food. But though we were as motionless as circumstances would permit, they never learned to trust us.

One—the mother, doubtless—did sometimes pay a flying visit to her three darlings under the leaves; but she undoubtedly felt that she took her life in her hands (so to speak), and it did not give her courage. She returned to her post and cried no less than before. We were not heartless; we could not bear to torture the timid creatures, and therefore we never stayed very long.

Every day we looked at the growing babies, who passed most of their time in sleep, as babies should; and at last came the time, sooner than expected, when we found the family had flitted. Nestlings cradled near the ground seem to be spared the long period in the nest endured by birdlings who must be able to fly before they can safely go. Young veeries and bobolinks, song sparrows and warblers, who build low, apparently take leave of the nursery as soon they can stand up. Thereafter the parents must seek them on the ground; and if the student follows their chirps, he will often see the droll little dumpy fellows running about or crouched under bushes until their wing feathers shall grow and lift them to the bird's world, above the dull earth.

After the exit of the family in the maples, we kept closer watch of the remaining nest. Every day we passed it, and not always at the same hour, yet never but once did we find the mother away, and seven days after that morning, when not one youngster had broken the shell, the family was gone.

The young birds in the maples we had seen in the nest for five days after they were hatched, so we were forced to believe that either the second nest had been robbed, or that the mother had watched for us, and flown to cover her babies after they were hatched, till we had paid our daily visit and passed on. This latter may be the correct conclusion, and if so, her conduct was entirely different from that of any veery I have seen.

Whatever cause had emptied the thrush cradle we found no signs of disturbance about it, and we heard no lamentations. But we did hear from every impenetrable tangle in the woods, the baby-cries of young thrushes; and we ventured to hope that no hawk or owl or squirrel, or other foe in feathers or in fur, had carried off the nestlings of that brave brown-eyed mamma.

XII. A MEADOW NEST

A bird's nest in the middle of a meadow is as isolated as if on an island; for the most eager bird student, though he may look and long afar off, will hesitate before he harrows the soul of the owner of the fair waving sea of grass by trampling it down. In such a secure place, among scattered old apple-trees, a pair of veeries had set up their household, surrounded and protected from every enemy who does not wear wings.

They were late in nesting, for young veeries were out everywhere. Doubtless the first home had been destroyed, and they had selected this retreat in the midst of the tall grass for its seclusion and apparent safety.

What dismay, then, must have filled the heart of the timid creatures when there arrived, one morning, a party of men and horses and machines, who proceeded at once, with the clatter and confusion which follows the doings of men, to lay low their green protecting walls, and expose their cherished treasures to the greed or the cruelty of their worst enemies! Not less their surprise and grief when, after the uproar of cutting, raking and carrying away their only screen, there entered the silent but watchful spies, who planted their stools in plain sight, to take note of all their doings.

The nest, with its babies three, was wide open to the sun; no one could pass without seeing it. It was in a cluster of shoots growing up from the roots of an old apple-tree, and so closely crowded between them that its shape was oval.

The nestlings were nearly ready to fly, and I hoped that birds brave enough to come out of the woods and build among apple-trees would be less afraid of people than the woods dwellers. So when I learned of my comrade's discovery I hastened at once to make the acquaintance of this, our fourth nesting-veery of the summer.

The parents were absent when I seated myself at some distance from their homestead to wait. They soon came, together, with food in their mouths; but their eager, happy manner vanished at sight of me, and they abandoned themselves to utter despair, after the manner of veeries. They stood motionless on neighboring perches, and cried and bewailed the anticipated fate of those babies for all of the short time that I was able to endure it. A kingbird came to the tree under which I sat, to see for himself the terrible bugaboo, and a robin or two, as usual, interested themselves in the affairs of a neighbor in trouble.

Thirty minutes proved to be as long as I could bring myself to stay, and then I meekly retired to the furthest corner of the field, where I made myself as inconspicuous as possible, and hoped I might be allowed to remain. Kingbird and robins accepted the compromise and returned to their own affairs; but the veeries by turns fed the babies and reviled me from a tree near my retreat, till I took pity on their distress and left the orchard altogether.

Not only does the veery exhibit this strong liking for solitude, and express the loneliness of the woods more perfectly than any other bird, with the exception, perhaps, of the wood-pewee; but his calls and cries are all plaintive, many of them sensational, and one or two really tragic.

His most common utterance, as he flits lightly from branch to branch, is a low, sweet "quee-o," sometimes hardly above a whisper. When everything is quiet about him one may often hear an extraordinary performance. Beginning the usual call of "quee-o," in a tender and mournful tone, he will repeat it again and again at short intervals, every time with more pathetic inflection, till the wrought-up listener cannot resist the feeling that the next sound must be a burst of tears. Although his notes seem melancholy to hearers, however, the beautiful bird himself is far from expressing that emotion in his manner.

Aside from the enchanting quality of his calls, and the thrilling magnetism of his song, the tawny thrush is an exceedingly interesting bird. In his reserved way he is socially inclined, showing no dislike to an acquaintance with his human neighbors, and even evincing a curiosity and willingness to be friendly, most winning to see.

Speak to one who, as you passed, has flown up from the ground and alighted on the lowest limb of a tree, looking at you with clear, calm eyes. He will not fly; he will even answer you. You may stand there for half an hour and talk to him and hear his replies. It seems as if it were the easiest thing in the world to inspire him with perfect confidence, to coax him to a real intimacy. But there is a limit to his trustfulness. When he has a nest and little ones to protect, as already shown, he is a different bird; he is wild with distress, and refuses to be comforted when one approaches the sacred spot.

This unfortunate distrust of one's intentions makes it very hard for a student who loves the individual bird to watch his nest. One can't endure to give pain to the gentle and winsome creature. The mournful, despairing cry of both parents, "ke-o-ik! ke-o-ik! ke-o-ik!" constantly repeated, makes me, at least, feel like a robber and a murderer, and no number of "facts" to be gained will compensate me for the suffering thus caused.

One more phase of veery character I was surprised and delighted to learn. Sitting on a log in the edge of the woods one evening, just at sunset, I listened to the singing of one of these birds quite close to me, but hidden from sight. I had never been so near a singer, and I was surprised to hear, after every repetition of his song, a low response, a sort of whispered "chee." Was it his mate answering, or criticizing his music? Was it the first note of his newly-fledged offspring? Or could it be *sotto voce* remarks of the bird himself? It was impossible to decide, and I went home much puzzled to account for it; but a day or two later the mystery was solved—the thrush showed himself to be a humorist.

The odd performance by which I discovered this fact I saw through my closed blind. The bird was in plain sight on a small dead tree, but it was a retired spot, where he was accustomed to see no one, and he evidently did not suspect that he had a listener.

He had eaten his fill from a cluster of elderberries I had hung on the tree, and he lingered to sing a little, as he often did. First he uttered a call, aloud, clear "quee-o," and followed it instantly by a mocking squawk in an undertone. I could hardly believe my eyes and ears, and at once gave much closer attention to him. As if for the express purpose of convincing me that I had not been mistaken, he instantly repeated his effort; and after doing so two or three times, he poured out his regular song in his sweet, ringing voice, and followed it by a whispered "mew," almost exactly in the tone of pussy herself.

He was not far from my window, across a small yard, and as plainly seen through my glass as though not six feet away. I saw his beak and throat, and am absolutely certain that he delivered every note. The absorbed singer stood there motionless a long time, and carried on this queer conversation with himself. It sounded precisely like two birds, one of whom was mocking or ridiculing the other in a low tone.

Sometimes the undertone, as said above, was a squawk; again it resembled a squeal; now it was petulant, as though the performer scoffed at his own singing; and then it was a perfect copy of the song itself, given in an indescribably sneering manner. I could think of nothing but the way in which one child will sometimes mock the words of another.

It was very droll, as well as exceedingly interesting, and I hope some day to study further this unfamiliar side of the thrush nature.

After my unsuccessful attempt to disarm the fears and suspicions of the meadow-nesting thrushes, we left the little family to its much loved solitude, and in a day or two the whole nestful departed.

XIII. A JUNE ROUND OF CALLS

"I should like to meet you two in that rig on Fifth Avenue," calmly said our hostess one morning in June, as we started out on our regular round of calls.

What a suggestion! We stared at each other with a new standard of criticism in our eyes. We were not exactly in ordinary visiting costume; but then, neither were we making ordinary visits, for the calling-list of June differs in every way from that of January. The neighbors at whose doors we appeared would be quite as well (or as ill) pleased to see us in our dull green woods dress, with fresh leaves on our hats to convey the impression that we were mere perambulating shrubs, with opera-glasses instead of cards, and camp-stools in place of a carriage, as though we had been in regulation array. Away we went, the big dog prancing ahead with the camp-stool of his mistress.

Our first call was upon a small dame very high up in the world, thirty feet at least. The mention of Fifth

Avenue suggests that possibly our manners were not above criticism. We introduced ourselves to Madam Wood-Pewee not by ringing and sending up cards, but by pausing before her door, seating ourselves on our stools, and leveling our glasses at her house. We felt, indeed, that we had almost a proprietary interest in that little lichen-covered nest resting snugly in a fork of a dead branch, for we had assisted in building it, at least by our daily presence, during the week or two that she spent in bringing, in the most desultory way, snips of material, fastening them in place, and moulding the whole by getting in the nest and pressing her breast against it, while turning slowly round and round. Now that she had really settled herself to sit, we never neglected to leave a card upon her, or so to speak, every morning.

As we approached we were pleased to see her trim lord and master bearing in his mouth what was no doubt intended for a delicate offering to cheer her weary hours, for a gauzy yellow wing stuck out on each side of his beak, suggesting something uncommonly nice within. He stood a moment till we should pass, looking the picture of unconsciousness, and defying us to assert that he had a house and home anywhere about that tree. But when we did not pass, after hesitatingly hopping from perch to perch nearer the nest, he deliberately diverted yellow wing from its original destiny, swallowed it himself, and wiped his beak with an air that said: "There now! What can you make out of that?"

Ashamed to have deprived the little sitter of her treat, we folded our stools and resumed our march.

How shall one put into words the delights of the woods in June without "dropping into poetry?" Does not our own native poet say:

"Who speeds to the woodland walks?
To birds and trees who talks?
Cæsar of his leafy Rome,
There the *poet* is at home."

But if one is not a poet, must he then suffer and enjoy in silence? When he puts aside the leafy portière and enters the cool green paradise of the trees, must he be dumb? Slowly, almost solemnly, we walked up the beautiful road with its carpet of dead leaves. It was as silent of man's ways as if he were not within a thousand miles, and we had all the enjoyment of the deep forest, with the comforting assurance that five minutes' walk would bring us to people.

A small family in dark slate-color and white, with a curious taste for the antique cave-dwelling, was next on our list. The home was an excavation in the soft earth, held together by the roots of an overturned tree, and everything was quiet when we arrived—the two well-grown infants sound asleep on their hair mattress. We sat down to wait, and in a moment we heard the anxious "pip" of the returning parents. They had been attending to their regular morning work, and both brought food for those youngsters, who woke inopportunely—as babies will—and demanded it instantly.

Junco—for he was the head of this household—paused on a twig near by, opened and shut his beautiful white-bordered tail, in the embarrassing consideration whether he should go in before our eyes and take the risk of our intentions, or let his evidently starving offspring suffer. He "eyed us over;" he waited till his modest little spouse, acting from feeling rather than from judgment (as was to be expected from one of her unreasoning sex), had slipped in from below, administered her morsel to those precious babies, and escaped unharmed. Then he plucked up courage, boldly entered his door, gave a poke behind it, and flew away.

A week later, after we had called as usual one morning and found the house empty, he brought his pretty snow-birdlings in their tidy striped bibs up to the grove at the back door, where we often heard his sharp trilling little song, and saw him working like some bigger papas to keep the dear clamorous mouths filled.

The Junco neighborhood was a populous part of our calling district. Behind his cave, in a high tree, lived a family of golden-winged woodpeckers, who "laughed" and talked as loud as they liked, scorning to look upon the two spies far below them. Not quite so self-possessed and bold were they a little later, when madam came up to the grass by the farmhouse with her young son to teach him to dig, for that is what she did. He was a canny youngster, though he was shy, and had no notion of being left in the lurch for a moment. If mamma flew to the fence, he instantly followed; did she return to the ground, baby was in a second at her side demanding attention. On one occasion while I was watching them behind my blind, the mother managed to slip away from him and disappear. In a moment he realized his deserted condition, stretched up, like a lost chicken, looking about on every side, and calling, in a plaintive tone, "pe-au! au!"

and then, "au! au! pe-au!" When at length he saw his mother, he burst into a loud cry of delight, and flew into a locust-tree, where I heard for a long time low complaining cries, as if he reproached her for leaving her baby alone on the fence.

On the right of the home of the golden wings, in a sapling not more than five feet from the ground, was the residence of a gay little redstart, which we had watched almost from the laying of the foundations. We made our visit. Yesterday there were two pearls of promise within; to-day, alas! nothing.

Squirrels, we said; for those beasts were the bugaboo of the woods to its feathered inhabitants. Hardly a nest was so high, so well hidden, or so closely watched, but some unlucky day a little fellow—sportsman, would you call him?— in gray or red fur, would find his chance, and make his breakfast on next year's song birds.

Musing on this and other tragedies among our friends, we silently turned to the next neighbor. At this door we could knock, and we always did. (We desired to be civil when circumstances permitted.) A rap or two on the dead trunk brought hastily to the door, twenty-five feet high, a small head, with a bright red cap and necktie, and eager, questioning eyes. Observing that he had guests, he came out, showing his black and white coat. But one glance was usually enough; he declined to entertain us, and instantly took his leave. We knew him well, however—the yellow-bellied woodpecker, or "sapsucker," as he was called in the vicinity. This morning we did not need to knock, for one of the family was already outside —a young woodpecker, clinging to the bark, and dressing up his nest-ruffled plumage for the grand performance, his first flight. We resolved at once to assist at the début, secured reserved seats with a good view, and seated ourselves to wait.

Didst ever, dear reader, sit in one position on a camp-stool without a back, with head thrown back, and eyes fixed upon one small bird thirty feet from the ground, afraid to move or turn your eyes, lest you miss what you are waiting for, while the sun moves steadily on till his hottest rays pour through some opening directly upon you; while mosquitoes sing about your ears (would that they sang only!), and flies buzz noisily before your face; while birds flit past, and strange notes sound from behind; while rustling in the dead leaves at your feet suggests snakes, and a crawling on your neck proclaims spiders? If you have not, you can never appreciate the enthusiasms of a bird student, nor realize what neck-breaks and other discomforts one will cheerfully endure to witness the first flight of a nestling.

This affair turned out, however, as in many another case of great expectations, to be no remarkable performance. When the débutante had made his toilet, he flew, as if he had done it all his life, to the next tree, where he began at once to call for refreshment, after his exertion.

Disappointed, we dropped our eyes, whisked away our insect tormentors, gathered up our properties, and passed on our way.

This was the farthest point of our wanderings. The way back was through a narrow path by the oven-bird's pretty domed nest, then between the tangle of wild-berry bushes and saplings, where a cuckoo had set up housekeeping, and where veeries and warblers had successfully hidden their nests, tantalizing us with calls and songs from morning till night; from thence through the garden, past the kitchen door, home.

XIV. A BOBOLINK RHAPSODY

Can anything be more lovely than a meadow in June, its tall grass overtopped by daisies, whose open faces,

> "Candid and simple and nothing-withholding and free,
> Publish themselves to the sky"?

One such I knew, despised of men as a meadow, no doubt, but glorious to the eye with its unbroken stretch of white bowing before the summer breeze like the waves of the sea, and charming as well to pewee and kingbird who hovered over it, ever and anon diving and bringing up food for the nestlings. When, to a meadow not so completely abandoned to daisies, where buttercups and red clover flourish among the grass, is added the music of the meadow's poet, the bobolink, surely nothing is lacking to its perfection.

Passing such a field one evening, I noted the babble of bobolinks, too far off to hear well, and the next day I set out down another path which passed through the meadow, to cultivate the acquaintance of the birds. It was a warm summer morning, near the middle of June, and when I reached the spot not a bobolink was in sight; but I sought a convenient bank under an old apple-tree, made myself as inconspicuous as possible, and waited. With these birds, however, as I soon found out, my precautions were unnecessary. They are not chary of their music; on the contrary, they appear to sing directly to a spectator, and they are too confident of the security of the nest to be disturbed about that. In a moment a black head with its buff cap appeared at the top of a grass stem, and instantly the black body, with its grotesque white decoration, followed.

The bird flew half a dozen feet, singing as he went, as if the movement of the wings set the music going, alighted a little nearer, sang again, and finally, concluding that here was something to be looked after, a human being, such as he was accustomed to see pass by, taking possession of a part of the bobolink domain, he flew boldly to a small tree a few yards from me. He alighted on the top twig, in plain sight, and proceeded to "look me over," a performance which I re-turned with interest. He was silent only a few seconds, but the sound that came from his beak amazed me; it was a "mew." If the cat-bird cry resembles that of a cat, this was a perfect copy of a kitten's weak wail. It was al-ways uttered twice in close succession, and sometimes followed by a harsh note that proclaimed his blackbird strain, a "chack!"

His utterance was thus: "mew, mew (quickly), chack!" and I interpreted it into a warning to me to leave the premises. I did not go, however, and after several repetitions his vigilance began to relax. He was really so full of sweet summer madness that it was impossible to keep up the role of stern guardian of the nests under the veil of buttercups and daisies, which he knew all the time I could never find. So, when he opened his mouth to say "chack," a note or two would irresistibly bubble out beside it, as if he said, "You really must go away, my big friend. We cannot have you in our fields; but, after all, isn't the morning delicious?"

After a long conflict between desire to sing and his conviction of duty as special policeman, which ludicrously suggested Mr. Dick in his struggle between longing to be foolish with David Copperfield and to be grave to please Miss Betsy, he fairly gave in and did sing—and such a burst! Everybody has tried his hand at characterizing this bird's incomparable song, but no one has fully expressed it, for words are not capable of it. Perhaps Mrs. Spofford has caught the spirit as well as any one:

"Last year methinks the bobolinks
Filled the low fields with vagrant tune,
The sweetest songs of sweetest June—
Wild spurts of frolic, always gladly
Bubbling, doubling, brightly troubling,
Bubbling rapturously, madly."

Expressing himself was so great a relief to my bobolink, after his unnatural gravity of demeanor, that he repeated the performance again and again. I say that he repeated it; I found that he had two ways of beginning, but after he got into his ecstasy I could think of nothing but how marvelous it was, so that whether the two differed all through I am not sure. It was every time a new rapture to me as well as to him. One of his beginnings that I had time to note before I was lost in the flood of melody was of two notes, the second a fifth higher than the first, with a "grace-note," very low indeed, before each one. The other beginning was also two notes, the second at least a fifth lower than the first, with an indescribable jerk between, and uttered so softly that if I had been a little further away I could not have heard it. It sounded like "tut, now."

Seeing that I remained motionless, the bird forgot altogether his uncongenial occupation of watchman, and launched himself into the air toward me, soaring round and round me, letting fall such a flood, such a torrent, of liquid notes that I thought half a dozen were singing—and then dropped into the grass. Soon others appeared here and there, and sang it mattered not how or where—soaring or beating the wings, on a grass stem, the top of a tree, hidden in the grass, or rudely rocked by the wind, they "sang and sang and sang."

Then for a while all was still. A turkey leading her fuzzy little brood about in the grass thrust her scrawny neck and anxious head above the daisies, said "quit! quit!" to me, and returned to the brooding mother-tones that kept her family around her. Tiring of my position while waiting for the concert to resume, I laid my head back among the ferns, letting the daisies and buttercups tower above my face—strangely enough, by this simple act realizing as never before the real motherhood of the earth.

While I lay musing, lo, a sudden burst of music above my head! A bobolink sailed over my face, not three feet from it, singing his merriest, and then dropped into the grass behind me. Oh, never did I so much wish for eyes in the back of my head! He must be almost within touch, yet I dared not move; doubtless I was under inspection by that keen dark eye, for the first movement sent him away with a whir.

My next visitors were a small flock of six or eight cedar-birds, who were seriously disturbed by my choice of a couch. Evidently the green tent above my head was their chosen tree, and they could not give it up. Finding me perfectly silent, they would come, perch in various parts of the branches, and turn their wise-looking black spectacles down to look at me, keeping up an animated conversation the while. We call the cedar-bird silent because he has, as generally supposed, but one low note; but he can put into that one an almost infinite variety of expressions. If I so much as moved a hand, instantly my Quaker-clad friends dived off the tree below the bank across the road, as if, in their despair, they had flung themselves madly into the brook at the bottom. But I did not suspect them of so rash an act, and, indeed, in a few minutes the apple-tree again resounded with their cries.

Meanwhile the sun marched relentlessly on, and the shadows without and the feelings within alike pointed to the dinner hour (12 p.m.). I rose, and thereby created a panic in my world. Six cedar-birds disappeared over the bank, a song sparrow flew shrieking across the field, a squirrel interrupted in his investigations fled madly along the rail fence, every few steps stopping an instant, with hindquarters laid flat and tail resting on the rail, to see if his head was still safe on his shoulders.

I gathered up my belongings and sauntered off toward home, musing, as I went, upon the bobolink family. I had not once seen or heard the little mates. Were they busy in the grass with bobolink babies? and did they enjoy the music as keenly as I did? How much I "wanted to know"! How I should like to see the nests and the nestlings! What sort of a father is the gay singer? (Some of the blackbird family are exemplary in this relation.) Does he drop his part of poet, of reveler of the meadows, I wonder, and come down to the sober prose of stuffing baby mouths? Are bobolinks always this jolly, delightful crowd? Are they never quarrelsome? Alas! it would take much more than one day, however sunny and however long, to tell all these things.

At the edge of the meadow I sat down again, hoping for one more song, and then came the crown of the whole morning, the choicest reserved for the last. A bird sailed out from behind the daisies, passed over my head, and delivered the most bewitching rhapsody I had yet heard. Not merely once did he honor me, but again and again without pausing, as if he intended to fill me as full of bobolink rapture as he was himself. His voice was peculiarly rich and full, and, what amazed me, his first three notes were an exact reproduction of the wood-thrush's (though more rapidly sung), including the marvelous organ-like quality of that bird's voice. I could have listened forever.

"Oh, what have I to do with time?
For this the day was made."

But when he had uttered his message he sank back into the grass, and I tore myself away from the bobolink meadow, and came home far richer and far happier than when I set out.

XV. THE BOBOLINK'S NEST

My acquaintance with the bobolink was resumed a year later in the lovely summer home of a friend in the Black River Country, within sight of the Adirondack hills. We had found many nests in the woods and orchards, but the meadow had been safe from our feet, partly because of the rich crops that covered it, but more, perhaps, because of the hopelessness of the search over the broad fields for anything so easily hidden as a ground nest.

One evening, however, our host with a triumphant air invited us to walk, declaring that he could show us a nest more interesting than we had found.

The gentleman was a joker, and his statements were apt to be somewhat embellished by his vivid imagination, so that we accepted them with caution; but now he looked exultant, and we believed him, especially as he took his hat and stick and started off.

Down the road we went, a single carriage-way between two banks of grass a yard high. After carefully taking his bearings by certain small elm-trees, and searching diligently about for an inconspicuous dead twig he had planted as a guide-post, our leader confidently waded into the green depths, parted the stalks in a certain spot, and bade us look.

We did. In a cosy cup, almost under our feet, were cuddled together three bird-babies.

"Bobolinks?" we cried in a breath.

"Yes, bobolinks," said our guide; "and you had to wait for an old half-blind man to find them for you."

We were too much delighted to be annoyed by his teasing; a bobolink's nest we never hoped to see.

Nor should we, but for a discovery of mine that very morning. Walking down that same road, I had noticed in the deep grass near the path a clump of exquisite wild flowers. They were of gorgeous coloring, shaded from deep orange to rich yellow, full petaled like an English daisy, and about the size of that flower, with the edge of every tiny petal cut in fairy-like fringe. I admired them for some minutes as they grew, and then gathered a handful to grace my room. As I came up to the house, my host stood on the steps; his eyes fell at once upon my nosegay, and a look of horror came into his face.

My heart sank. Had I unwittingly picked some of his special treasures, some rare exotic which he had cultivated with care?

"Where did you find that stuff?" he demanded. I was instantly relieved; no man will call a treasure "stuff."

"In the meadow," I answered. "What is it?"

"You must show me the exact spot," he said, emphatically. "I shall have a man out at once, to get it up, root and branch. It's the devil's paintbrush."

"Then his majesty has good taste in color," I said.

"That stuff," he went on, "spreads like wildfire. It'll eat up my meadow in a year."

I turned back and showed him the spot from which my flowers had come, pointing out at the same time two or three other clumps I could see farther out in the waving green sea, and before long his farmer and he were very busy over them.

Now it appeared that in tramping about the deep grass, where we bird-students dared not set our feet, he had nearly stepped on a bobolink, who flew, and thus pointed out her nest; and he had taken its bearings with the intention of putting us to shame.

We looked long at the tiny trio so compactly packed in their cradle, till they awoke and demanded supplies. Then we carefully replanted the dead stick, taking its exact bearings between three trees, drew a few grass-stems together in a braid at the margin so that we should not lose what we had so accidentally gained, and then we left them.

During this inspection of the nest, the "poet of the year" and his spouse were perched on two neighboring trees, utterly unmoved by our movements. They were, no doubt, so perfectly confident of the security of the hiding-place that it never occurred to them even to look to see what we three giants were doing. At least, such we judged were their sentiments by the change in their manners somewhat later, when they thought we were likely to make discoveries.

The meadow itself had been our delight for weeks. When we arrived, in the beginning of June, it was covered with luxuriant clumps of blue violets, and great bunches of blue-eyed grass that one might gather by the handful at one picking. Later the higher parts were thickly sprinkled with white where

"Gracefully as does the fawn,
Sweet Marguerites their dainty heads uphold,"

while the hollows were golden with buttercups. Then the grass under the warm June sun stretched up inch by inch till it was three or four feet high and very thick. Meanwhile a bobolink or two, and as many meadow-larks had taken possession of it, and it was made still richer by the sweet minor strains of the lark, and the song of the bird who,

"like the soul
Of the sweet season vocal in a bird,
Gurgles in ecstasy we know not what."

The evening after our humiliation—which we lost sight of in our joy—we returned to the charmed spot, parted again the sweet grass curtains and gazed down at the baby bobolinks, while the parents perched on two trees as before and paid not the smallest attention to us.

We passed on down the road to the gate where we could look into a neighboring pasture and watch for a pair of red-headed woodpeckers who lived in that pleasant place, and catch the reflection of the sunset in the northern sky. While we lingered there, I looked with my glass back at the bobolinks, and chanced to see Bobby himself in the act of diving into the grass. When he came out he seemed to notice me, and instantly began trying to mislead me.

He came up boldly, flew to another spot where a weed lifted its head above the green, and dropped into the grass exactly as though he was going to the nest; then he rose again, repeated his tactics, pausing every time he came out and calling, as if to say, "This is my home; if you're looking for a nest, here it is!" His air was so business-like that it would naturally deceive one not possessed of our precious secret, the real spot where his three babies were cradled, and one might easily be led all over the meadow by the wily fellow.

For six successive days we paid our short visits, and found the nestlings safe. They did not seem to mature very fast, though they came to look up at us, and open their mouths for food. But on the seventh day there was a change in Master Robert's behavior. On the afternoon of this day, wishing to observe their habits more closely, I found a seat under a tree at some distance, not near enough, as I thought, to disturb them.

I did disturb them sorely, however, as instantly appeared. The calmness they had shown during all the days we had been looking at the nest was gone, and they began to scold at once. The head of the family berated me from the top of a grass-stem, and then flew to a tall old stump, and put me under the closest surveillance, constantly uttering a queer call like "Chack-que-dle-la," jerking wings and tail, and in every way showing that he considered me rudely intrusive and altogether too much interested in his family affairs. I admitted the charge, I could not deny it; but I did not retire.

At last he apparently determined to insist upon my going, for he started from his high perch directly toward me. Swiftly and with all his force he flew, and about twenty feet from me swooped down so that I thought he would certainly strike my face. I instinctively dodged, and he passed over, so near that the wind from his wings fanned my face. This was a hint I could not refuse to take. I left him, for the time.

That evening when we went for our usual call, lo! the nest was empty. At not more than seven or eight days of age, those precocious infants had started out in the world! That explained the conduct of the anxious papa in the afternoon, and I forgave him on the spot. I understood his fear that I should discover or step on his babies three, scattered and scrambling about under all that depth of grass. The abandoned homestead, which we carefully examined, proved to be merely a cup-shaped hollow in the ground, slightly protected by a thin lining.

In a few days the wandering younglings were up in front of the house, where we could watch the parents drop into the grass with food; and where, of course, they were safe from anybody's intrusion. I had one more encounter with his lordship. After the young had been out a week or more, they seemed in their moving about to get back near to the old place. As I took my usual walk one evening, down the carriage drive to the gate, I found two pairs of bobolinks on one tree; the two mothers with food in their mouths, evidently intended for somebody down in the grass; and the two fathers, very much disturbed at my appearance. They greeted me with severe and reproving "chacks," and finally favored me with the most musical call I have heard from the sweet-voiced bird of the meadow. It was like "kee-lee!" in loud and rich tones, and it was many times repeated.

I assured them that I had no wish to disturb their little ones; though, if I had been able to lift the whole grassy cover to peep at the two small families hidden there, I fear I should have yielded to the temptation.

Our bird had been somewhat erratic in making his home far from his fellows—so social are these birds even in nesting-time; but now he was joined by more of his kind from the meadows below, and to the beautiful waving carpet of green, dotted here and there with great bunches of black-eyed Susans and devil's paint-brushes (what names!), and sprinkled all over with daisies, now beginning to look a little disheveled and wild, was added the tantalizing interest of dozens of little folk running about under its shelter.

The next week brought to the meadow what must seem from the bobolink point of view almost the end of the world. Men and horses and great rattling machines, armed with sharp knives, which laid low every stem of grass and flower, and let the light of the sun in upon the haunts and the nests of the bobolink babies.

Happily, however, not all the earth is meadow and subject to this annual catastrophe; and I think the whole flock took refuge in a pasture where they were safe from the hay-cutters, and had for neighbors only the cows and the crow babies.

XVI. THE TANAGER'S NEST

One of the prettiest memory-pictures of my delightful June on the banks of the Black River is the nest of a scarlet tanager, placed as the keystone of one of Nature's exquisite living arches. The path which led to it was almost as charming as the nest itself. Lifting a low-hanging branch of maple at the entrance to the woods, we took leave of the world and all its affairs, and stepped at once into a secluded path. Though so near the house, the woods were solitary, for they were private and very carefully protected.

Passing up the rustic foot-path, under interlacing boughs of maple and beech, we came at length to a sunny open spot, where all winter grain is kept for partridges, squirrels, and other pensioners who may choose to come. From this little opening one road turned to the wild-berry field, where lived the cuckoo and the warblers; another opened an inviting way into the deep woods; a third went through the fernery. We took that, and passed on through a second lovely bit of wood, where the ground was wet, and ferns of many kinds grew luxuriantly, and the walk was mostly over a dainty corduroy of minute moss-covered logs.

At the end of the fernery are two ways. The first runs along the edge of the forest, whose outlying saplings hang over and make a cool covered walk. Down this path I almost had an adventure one day. The morning was warm and I was alone. As I came out of this covered passage, beside an old stump, I noticed in a depression in the ground at my feet a squirming mass of fur. On looking closer I saw four or five little beasts rolling and scrambling over each other. They were as big, perhaps, as a month-old kitten, but they were a good deal more knowing than pussy's babies, for as I drew near they stopped their play and waited to see what would happen. I looked at them with eager interest. They were really beautiful; black and white in stripes, with long bushy tails. Black and white, and so self-possessed!—a thought struck me. "Mephitis," I gasped, and instantly put several feet more between us. So attractive and playful were they, however, that notwithstanding I feared it might be hard to convince their mamma, should she appear, of my amiable intentions, I could not resist another look.

Calm as a summer morning walked off one of the mephitis babies, holding his pretty tail straight up like a kitten's, while the other four went on with their frolic in the grass. At this moment I heard a rustle in the dead leaves, and having no desire to meet their grown-up relatives, I left in so great haste that I took the wrong path, and finally lost myself for a time in a tangle of wild raspberry bushes, whose long arms reached out on every side to scratch the face and hands or catch the dress of the unwary passer-by.

The other of the two ways spoken of was a road, soft-carpeted with dead leaves. To reach the tanager's nest we took that, and came, a little further on, to a big log half covered with growing fungi and laid squarely across the passage. This was the fungus log, another landmark for the wanderer unfamiliar with these winding ways. On this, if I were alone, I always rested awhile to get completely into the woods spirit, for this is the heart of the woods, with nothing to be seen on any side but trees. Cheerful, pleasant woods they are, of sunny beech, birch, maple, and butternut, with branches high above our heads, and a far outlook under the trees in every direction.

There is no gloom such as evergreens make; no barricade of dark impenetrable foliage, behind which might lurk anything one chose to imagine, from a grizzly bear to an equally unwelcome tramp.In this lovely spot come together four roads and a path, and to the pilgrim from cities they seem like paths into paradise. That on the right leads by a roundabout way to the "corner," where one may see the sunset. The next, straight in front, is the passage to the nest of the winter wren. The far left invites one to a wild tangle of fallen trees and undergrowth, where veeries sing, and enchanting but maddening warblers lure the bird-lover on, to scramble over logs, wade into swamps, push through chaotic masses of branches, and, while using both hands to make her way, incidentally offer herself a victim to the thirsty inhabitants whose stronghold it is. All this in a vain search for some atom of a bird that doubtless sits through the whole, calmly perched on the topmost twig of the tallest tree, shielded by a leaf, and pours out the tantalizing trill that draws one like a magnet.

Between this road and the wren's highway a path runs upward. It is narrow, and guarded at the opening by a mossy log to be stepped over, but it is most alluring. Up that route we go. On the left as we pass we notice two beautiful nests in saplings, so low that we can look in; redstarts both, and nearly always we find madam at home. We pass on, step over a second mossy log, pause a moment to glance at a vireo's hanging cradle on the right, and arrive at length at a crossing road, on the other side of which our path goes on, with a pile of logs like a stile to go over. Over the logs we step, walk a rod or two further, stop beside the blackened trunk of a fallen tree, turn our faces to the left, and behold the nest.

Before us is one of nature's arches. A maple sapling, perhaps fifteen feet high, has in some way been bowed till its top touched the ground and became fastened there, a thing often seen in these woods. Thus diverted from its original destiny of growing into a tree, it has kept its "sweetness and light," sent out leaves and twigs through all its length, and become one of the most beautiful things in the woods—a living arch. Just in the middle of this exquisite bow, five feet above the ground, is the tanager's nest, well shielded by leaves.

We never should have found it if the little fellow in scarlet had not made so much objection to our going up this particular passage that we suspected him of having a secret in this quarter. He went ahead of us from tree to tree, keeping an eye on us, and calling, warily, "chip-chur!" When we sat down a few moments to see what all the fuss was about, we saw his spouse in her modest dress of olive green on a low branch. She, too, uttered the cry "chip-chur!" and seemed disturbed by our call. Looking around for the object of their solicitude, our eyes fell at the same instant on the nest. We dared not speak, but an ecstatic glance from my comrade, with a hand laid on her heart to indicate her emotions, announced that our hopes were fulfilled; it was the nest we were seeking.

The birds, seeing that we meant to stay, flew away after a while, and we hastened to secrete ourselves before they should return, by placing our camp-stools in a thick growth of saplings just higher than our heads. We crowned ourselves with fresh leaves, not as conquerors, but as a disguise to hide our heads. We daubed our faces here and there with an odorous (not to say odious) preparation warranted to discourage too great familiarity on the part of the residents already established in that spot. We subsided into silence.

The birds returned, but were still wary. As before, the male perched high and kept a sharp eye out on the country around, and I have no doubt soon espied us in our retreat. Madam again tried to "screw her courage up" to visit that nest. Nearer and nearer she came, pausing at every step, looking around and calling to her mate to make sure he was near. At last, just as she seemed about to take the last step and go in, and we were waiting breathless for her to do it, a terrific sound broke the silence. The big dog, protector and constant companion of my fellow-student, overcome by the torment of mosquitoes, and having no curiosity about tanagers to make him endure them, had yielded to his emotions and sneezed. Away went the tanager family, and, laughing at the absurd accident, away we went too, happy at having discovered the nest, and planning to come the next day. We came next day, and many days thereafter, but never again did we see the birds near. They abandoned the nest, doubtless feeling that they had been driven away by a convulsion of nature.

One day, somewhat later, in the winter wren's quarter, where there were pools left by a heavy rain, we met them again. Madam was bathing, and her husband accompanied her as guard and protector. They flew away together. All of June we heard him sing, and we often followed him, but never again did we surprise a secret of his, till the very last day of the month. We had been making a visit to our veery nests, and on our way back noticed that the tanager was more than usually interested in our doings. He seemed very busy too, with the air of a person of family. While we were watching to see what it meant, he caught a flying insect and held it in his mouth. Then we knew he had little folk to feed, so we seated ourselves on the fungus log, and waited for him to point one out. He did. He could not resist giving that delicate morsel to his first-born.

With many wary approaches, he dropped at last into the scanty undergrowth, and there, a foot above the ground, we saw the young tanager. He was a little dumpling of a fellow, with no hint in his baby-suit of the glory that shall clothe him by and by. But where was the mother? and where had they nested? But for that untimely sneeze, as I shall always believe, they would have made their home in that beautiful nest on the arch, and we should have been there to see.

XVII. THE WILES OF A WARBLER

"Hark to that petulant chirp! What ails the warbler?
Mark his capricious ways to draw the eye."

We called him the blue, but that was not his whole name by any means. Fancy a scientist with a new bird to label, contenting himself with one word! His whole name is—or was till lately—black-throated blue-backed warbler, or *Dendroica cœrulescens*, and that being fairly set down for future reference for whom it may concern, I shall call him henceforth, as we did in the woods, the blue.

For a day or two at first he was to us, like many another of his size, only a "wandering voice." But it was an enticing voice, a sweet-toned succession of *z-z-z* in ascending scale, and it was so persistent that when we really made the attempt, we had no trouble in getting sight of the little beauty hardly bigger than one's thumb. He was a wary little sprite, and though he looked down upon us as we turned opera-glasses toward him—a battery that puts some birds into a panic—he was not alarmed. He probably made up his mind then and there, that it should be his special business to keep us away from his nest, for really that seemed to be his occupation. No sooner did we set foot in the woods than his sweet song attracted us. We followed it, and he, carelessly as it seemed, but surely, led us on around and around, always in a circle without end.

My fellow bird-student became fairly bewitched, and could not rest till she found his nest. For my part I gave up the warbler family long ago, as too small, too uneasy, too fond of tree-tops, to waste time and patience over. In these her native woods, my comrade led in our walks, and the moment we heard his tantalizing *z-z-z* she turned irresistibly toward it. I followed, of course, happy to be anywhere under these trees.

One morning she tracked him inch by inch till she was fortunate enough to trace him to a wild corner in the woods given up to a tangle of fallen trees, saplings, and other growth. She went home happy, sure she was on the trail. The next day we turned our steps to that quarter and penetrated the jungle till we reached a moderately clear spot facing an impenetrable mass of low saplings. There we took our places, to wait with what patience we might for the blue.

Our lucky star was in the ascendant that day, for we had not been there three minutes before a small, inconspicuous bird dropped into the bushes a few feet from us. My friend's eye followed her, and in a second fell upon the nest the little creature was lining, in a low maple about two feet from the ground.

But who was she? For it is one of the difficulties about nests, that the brightly-colored male, whom one knows so well, is very sure not to show himself in the neighborhood, and his spouse is certain to look just like everybody else. However, there is always some mark by which we may know, and as soon as the watcher secured a good look she announced in an excited whisper, "We have it! a female blue, building."

So it proved to be, and we planted our seats against trees for backs, and arranged ourselves to stay. The dog seeing this preparation, and recognizing it as somewhat permanent, with a heavy sigh laid himself out full length, and composed himself to sleep.

The work over that nest was one of the prettiest bits of bird-life I ever watched. Never was a scrap of a warbler, a mere pinch of feathers, so perfectly delighted with anything as she with that dear little homestead of hers. It was pretty; it looked outside like the dainty hanging cradle of a vireo, but instead of being suspended from a horizontal forked twig, it was held in an upright fork made by four twigs of the sapling.

The little creature's body seemed too small to hold her joy; she simply could not bring her mind to leave it. She rushed off a short distance and brought some infinitesimal atom of something not visible to our coarse sight, but very important in hers. This she arranged carefully, then slipped into the nest and moulded it into place by pressing her breast against it and turning around.

Thus she worked for some time, and then seemed to feel that her task was over, at least for the moment. Yet she could not tear herself away. She flew six inches away, then instantly came back and got into the nest, trying it this way and that. Then she ran up a stem, and in a moment down again, trying that nest in a new way, from a fresh point of view. This performance went on a long time, and we found it as impossible to leave as she did; we were as completely charmed with her ingenuous and bewitching manners as she was with her new home.

Well indeed was it that we stayed that morning and enriched ourselves with the beautiful picture of bird ways, for like many another fair promise of the summer it came to naught.

We had not startled her, she had not observed us at all, nor been in the least degree hindered in her work by our silent presence, twenty feet away and half hidden by her leafy screen. But the next day she was not there. After we had waited half an hour, my friend could no longer resist a siren voice that had lured us for days (and was never traced home, by the way). I offered to wait for the little blue while she sought her charmer.

We were near the edge of the woods, and she was obliged to pass through part of a pasture where sheep were kept. Her one terror about her big dog was that he should take to making himself disagreeable among sheep, when she knew his days would be numbered, so she told him to stay with me. He had risen when she started, and he looked a little dubious, but sat down again, and she went away.

He watched her so long as she could be seen and then turned to me for comfort. He came close and laid his big head on my lap to be petted. I patted his head and praised him a while, and then wished to be relieved. But flattery was sweet to his ears, and the touch of a hand to his brow—he declined to be put away; on the contrary he demanded constant repetition of the agreeable sensations. If I stopped, he laid his heavy head across my arm, in a way most uncomfortable to one not accustomed to dogs. These methods not availing, he sat up close beside me, when he came nearly to my shoulder and leaned heavily against me, his head resting against my arm in a most sentimental attitude.

At last finding that I would not be coaxed or forced into devoting myself wholly to his entertainment, he rose with dignity, and walked off in the direction his mistress had gone, paying no more attention to my commands or my coaxings than if I did not exist. If I would not do what he wished, and pay the price of his society, he would not do what I asked. I was, therefore, left alone.

I was perfectly quiet. My dress was a dull woods tint, carefully selected to be inconspicuous, and I was motionless. No little dame appeared, but I soon became aware of the pleasing sound of the blue himself. It drew nearer, and suddenly ceased. Cautiously, without moving, I looked up. My eyes fell upon the little beauty peering down upon me. I scarcely breathed while he came nearer, at last directly over my head, silent, and plainly studying me. I shall always think his conclusion was unfavorable, that he decided I was dangerous; and I, who never lay a finger on an egg or a nest in use, had to suffer for the depredations of the race to which I belong. The pretty nest so doted upon by its little builder was never occupied, and the winsome song of the warbler came from another part of the wood.

We found him, indeed, so often near this particular place, a worse tangle, if possible, than the other, that we suspected they had set up their household gods here. Many times did my friend and her dog work their way through it, while I waited outside, and considered the admirable tactics of the wary warbler. The search was without result.

Weeks passed, but though other birds interested us, and filled our days with pleasure, my comrade never ceased longing to find the elusive nest of that blue warbler, and our revenge came at last. Nests may be deserted, little brown spouses may be hidden under green leaves, homesteads may be so cunningly placed that one cannot find them, but baby birds cannot be concealed. They will speak for themselves; they will get out of the nest before they can fly; they will scramble about, careless of being seen; and such is the devotion of parents that they must and will follow all these vagaries, and thus give their precious secret to whoever has eyes to see.

One day I came alone into the woods, and as I reached a certain place, sauntering along in perfect silence, I evidently surprised somebody, for I was saluted by low "smacks" and I caught glimpses of two birds who dived into the jewel-weed and disappeared. A moment later I saw the blue take flight a little farther off, and soon his song burst out, calm and sweet as though he had never been surprised in his life.

I walked slowly on up the road, for this was one of the most enchanting spots in the woods, to birds as well as to bird-lovers. Here the cuckoo hid her brood till they could fly. In this retired corner the tawny thrush built her nest, and the hermit filled its aisles with music, while on the trespass notices hung here, the yellow-bellied woodpecker drummed and signaled. It was filled with interest and with pleasant memories, and I lingered here for some time.

Then as the road led me still farther away, I turned back. Coming quietly, again I surprised the blue family and was greeted in the same manner as before. They had slipped back in silence during my absence, and the young blues were, doubtless, at that moment running about under the weeds.

Thus we found our warbler, the head of a family, hard at work as any sparrow, feeding a beloved, but somewhat scraggy looking, youngster, the feeble likeness of himself. There, too, we found the little brown mamma, the same, as we suppose, whose nest-building we had watched with so much interest. She also had a youngster under her charge. But how was this! a brown baby clad like herself! Could it be that the sons and daughters of this warbler family outrage all precedent by wearing their grown-up dress in the cradle? We consulted the authorities and found our conclusion was correct.

Henceforth we watched with greater interest than before. Every day that we came into the woods we saw the little party of four, always near together, scrambling about under the saplings or among the jewel-weed, or running over the tangled branches of a fallen tree, the two younger calling in sharp little voices for food, and the elders bustling about on low trees to find it.

We soon noticed that there was favoritism in the family. Papa fed only the little man, while mamma fed the little maid, though she too sometimes stuffed a morsel into the mouth of her son. Let us hope that by this arrangement both babies are equally fed, and not, as is often the case, the most greedy secures the greatest amount of food.

We had now reached the last of July, and the woods were full of new voices, not alone the peeps or chirps of birdlings impatient for food. There were baffling rustles of leaves in the tree-tops, rebounds of twigs as some small form left them, flits of strange-colored wings—migration had begun. Now, if the bird-student wishes not to go mad with problems she cannot solve, she will be wise to fold her camp-stool and return to the haunts of the squawking English sparrow and the tireless canary, the loud-voiced parrot, and the sleep-destroying mockingbird. I did.

XVIII. A RAINY-DAY TRAMP

Before I opened my eyes in the morning I knew something had happened, for I missed the usual charm of dawn. A robin, to be sure, made an effort to lead, as was his custom, and sang out bravely once or twice; a song sparrow, too, flitted into the evergreen beside my window, and uttered his sweet and cheery little greeting to whom it might concern. But those were the only ones out of the fourteen voices we were accustomed to hear in the morning.

When I came out on the veranda not a note was to be heard and not a bird seen excepting a woodpecker, who bounded gayly up the trunk of a maple, as if sunshine were not essential to happiness, and a chipping-sparrow, who went about through the dripping grass with perfect indifference to weather, squabbling with his fellow-chippies, and picking up his breakfast as usual.

I seated myself in the big rocker, and turned toward the woods, a few rods away. The rain, which had fallen heavily for hours, light and fine now, drew a shimmering veil before the trees—a veil like a Japanese bead-hanging, which hides nothing, only the rain veil was more diaphanous than anything fashioned by human hands. It did not conceal, but enhanced the charm of everything behind it, lending a glamour that turned the woods into enchanted land.

Before the house how the prospect was changed! The hills and Adirondack woods in the distance were cut sharply off, and our little world stood alone, closed in by heavy walls of mist.

My glass transported me to the edge of the side lawn, where I looked far under the trees, and rejoiced in the joy of the woods in rain. The trees were still, as if in ecstasy "too deep for smiling;" the ferns gently waved and nodded. Every tiny leaf that had thrust its head up through the mould, ambitious to be an ash or a maple or a fern, straightened itself with fullness of fresh life. The woods were never so fascinating, nor showed so plainly

"The immortal gladness of inanimate things."

A summer shower the birds, and we, have reason to expect, and even to enjoy, but a downpour of several hours, a storm that lays the deep grass flat, beats down branches, and turns every hollow into a lake, was more than they had provided for, I fear. My heart went out to the dozens of bobolink and song-sparrow babies buried under the matted grass, the little tawny thrushes wandering around cold and comfortless on the soaked ground in the woods, the warbler infants—redstart and chestnut-sided—that I knew were sitting humped up and miserable in some watery place under the berry bushes, the young tanager only just out of the nest, and the two cuckoo babies, thrust out of their home at the untimely age of seven days, to shiver around on their weak blue legs.

My only comfort was in thinking of woodpecker little folk, the yellow-bellied family whose loud and insistent baby cries we had listened to for days, the downy and hairy, and the golden-wing. They were all warm and snug, if they could only be persuaded to stay at home. But from what I have seen of young birds, when their hour strikes they go, be it fair or foul. To take the bitter with the sweet is their fate, and no rain, however driving, no wind, however rough, can detain them an hour when they feel the call of the inner voice which bids them go. I have seen many birdlings start out in weather that from our point of view should make the feathered folk, old or young, hug the nest or any shelter they can find.

In the afternoon the rain had ceased, and we went out. How beautiful we found the woods! More than ever I despair of

"Putting my woods in song."

Every fresh condition of light brings out new features. They are not the same in the morning and the afternoon; sunshine makes them very different from a gray sky; and heavy rain, which hangs still in drops from every leaf and twig, changes them still more.

This time the tree-trunks were the most noticeable feature. Thoreau speaks of rain waking the lichens into life, and we saw this as never before. Not only does it bring out the colors and give a brightness and richness they show at no other time, but it raises the leaves—if one may so call them—makes them stand out fresh. The beeches were marvelous with many shades of green, and of pink, from a delicate blush over the whole tree, to bright vermilion in small patches. The birches, "most shy and ladylike of trees," were intensely yellow; some lovely with dabs of green, while others looked like rugged old heroes of many battles, with great patches of black, and ragged ends of loosened bark fringing them like an Indian's war dress, up to the branches. Every hollow under the trees had become a clear pond to reflect these beauties, and lively little brooks rippled across the path, adding to the woods the only thing they lacked—running water.

Instinctively our feet turned up the path to the oven-bird's nest, so narrow that we brushed a shower from every bush. There he was, singing at that moment. "Teacher! teacher! teacher!" he called, with head thrown up and wings drooped. And then while we looked he left his perch, and passed up between the branches out of our sight, his sweet ecstatic love-song floating down to delight our souls.

Surely, we thought, all must be well in the cabin among the dead leaves, or he could not sing so. Yet life had not been all rose-colored to the little dame whom we had surprised several days before, bringing great pieces of what appeared to be lace, to line the nest she had made so wonderfully. We had watched her, breathless, for a long time, while she went back and forth carrying in old leaves, softened, bleached, and turned to lace by long exposure, arranged each one carefully and moulded it to place by pressing her breast against it, and turning round and round in the nest. Curious enough she looked as she alighted at some distance away, and walked—not hopped—to her little "oven," holding the almost skeletonized leaf before her like an apron, so busy that she did not observe that she had visitors.Then came a day when, on reaching our usual place, we found that an accident had happened. The dainty roof was crushed in, and the poor little egg, for which such loving preparations had been made, lay pathetically on the ground outside the door. My comrade crept carefully up, raised the tiny roof to place, and with deft fingers put a twig under as a prop to hold it, then gently laid the pretty egg in the lace-lined nest.

The next day we hurried out to see if the bird had resented our clumsy human help. But no; like the wise little creature she was, she had accepted the goods the gods had provided, and laid a second pearl beside the first.

On our next visit, therefore—especially when we heard the gleeful song of her (supposed) mate—we came up with confidence to see our little oven-bird homestead. But, alas! somebody not so loving as we had been there; the two pretty eggs were gone, not a sign of them to be seen, and the nest was deserted. Yet we could not give up a hope that she would return, and day after day our steps turned of themselves to the oven-bird's nook. This rainy day, as a dozen times before, we found the little house still empty, and as before we turned sadly away, when suddenly a new sound broke the stillness. "Wuk! wuk! wuk! wa-a-a-ah! wa-a-a-ah!" it cried. It was the exact tone of a young baby, a naive and innocent cry.

What could it be? Was some tramp mother hidden behind the bushes? Was it a new bird with this unbird-like cry? I was startled. But my friend was smiling at my dismay. She pointed to the crotch of a tree, and there a saucy gray squirrel lay sprawled out flat, uttering his sentiments in this abominable parody on the human baby cry. I believe the first squirrel learned it from some deserted infant, and handed it down as a choice joke upon us all. At any rate this performer was not suffering as his tones would indicate; for seeing that he had an audience more interested than he desired, he pulled himself together, whisked his bushy tail in our faces, and disappeared behind the trunk, from whence, in one instant, his head was thrust on one side and his tail on the other. And so he remained as long as we were in sight.

This absurd episode changed our mood, and soon we tramped gayly back over the soft leaf-covered paths, fording the newly formed brooks, shaking showers upon ourselves from the saplings, and arriving at last, dripping but happy, on the veranda, where, after donning drier costumes, we spent the rest of the day watching the birds that came to the trees on the lawn.

XIX. THE VAGARIES OF A WARBLER

The bird lover who carries a glass but never a gun, who observes but never shoots, sees many queer things not set down in the books; freaks and notions and curious fancies on the part of the feathered folk, which reveal an individuality of character as marked in a three-inch warbler as in a six-foot man. Some of the idiosyncrasies of our "little brothers" may be understood and explained from the human standpoint, others are as baffling as "the lady, or the tiger?"

One lovely and lazy day last July—the fourth it was—a perfect day with not a cannon nor even a cracker to disturb its peace, my comrade and I turned our steps toward the woods, as we had for the thirty-and-three mornings preceding that one.

This morning, however, was distinguished by the fact that we had a special object. In general, our passage through the woods was an open-eyed (and open-minded) loitering walk, alternated with periods of rest on our camp-stools, wherever we found anything of interest to detain us.

On this Fourth of July we were in search of a warbler—one of the most tantalizing, maddening pursuits a sensible human being can engage in. Fancy the difficulty of dragging one's self, not to mention the flying gown, camp-stool, opera-glass, note-book and other impedimenta through brush and brier, over logs, under fallen trees, in the swamp and through the tangle, to follow the eccentric movements of a scrap of a bird the size of one's finger, who proceeds by wings and not by feet, who goes over and not through all this growth.

The corner to which we had traced our "black-throated blue," and where we suspected he had a nest, presented a little worse than the usual snarl of saplings and fallen branches and other hindrances, and the morning was warm. My heart failed me; and as my leader turned from the path I deserted. "You go in, if you like," I said; "I'll wait for you here."

I seated myself, and she went on. For a few minutes I heard the cracking of twigs, the rustle of her movements against the bushes, the heavy tread of her big dog, and then all was silent.

It was—did I say it was a fair morning?—not a breath of air was stirring. My seat was in a rather open spot at the foot of a big butternut tree; and I could look far up where its branches spread out wide and held their graceful leafy stars against the blue.

In the woods I am never lonely; but I was not this morning alone. Near by a vireo kept up his tireless song;

a gray squirrel peeped curiously at me from behind a trunk, his head showing on one side and his tail on the other; an oven-bird stole up behind to see what manner of creature this was, and far off I could hear the tanager singing.

I did not notice the time; but after a while I became conscious of a low whistle which seemed to mingle with my reveries, and might have been going on for some minutes. Suddenly it struck me that it was the call of my fellow-student, and I started up the road wondering lazily if she had found the nest, and, to tell the truth, not caring much whether she had or not. For, to tell the whole truth, I had long ago steeled my heart against the fascinations of those bewitching little sprites who never stay two seconds in one spot, and sternly resolved never, *never* to get interested in a warbler.

My companion, however, was not so philosophical or so cool. She never could withstand the flit of a warbler wing; she would follow for half a day the absurd but enchanting little trill; and she regularly went mad (so to speak) at every migration, over the hundred or two, more or less, varieties that made this wood a resting-place on their way. Now, I could resist the birds by never looking at them, but I could not resist my friend's enthusiasm; so when she started on a warbler trail, I generally followed, as a matter of course. And I admit that the blue, to which we shortened his name, was a beauty and a charming singer.

I passed quietly up the road toward the continued low calls, and soon saw the student, not far from the path, in a clearer spot than usual, sitting against a maple sapling, with her four-footed protector at her feet. When

I came in sight she beckoned eagerly but silently, and I knew she had found something; probably the nest, I thought. As quietly as might be under the circumstances (namely, a passage through dead leaves, brittle twigs, unexpected hollows, etc.), I crept to her side, planted my camp-stool near hers, and sat down, in obedience to her imperious gesture.

"Now look," she whispered, pointing to a nest in plain sight.

"Why that's the redstart nest we saw yesterday from the road," I answered in the same tone, somewhat disappointed, it must be said, for redstart nests were on about every third sapling in the woods."Yes; but see what's going on," she added, excitedly.

"I see," I replied; "there is a young bird on the edge of the nest and its mother is feeding it;" and I was about to lower my glass and ask what there was surprising about that, when she went on:

"Keep looking! There! Who's that?"

"Why that's—why—that's a chestnut-sided warbler! and—what?—he feeds the same baby!" I gasped, interested now as much as she.

"There!" she exclaimed, triumphantly, "I wanted you to see that with your own eyes, since you scorn to look at the warblers. He has been doing that ever since I left you. I couldn't bear to let him out of my sight!"

At that moment the warbler appeared again, and the wise redstart baby, who at least knew enough to take a good thing when it offered, opened his ever-ready mouth for the bit of a worm he brought.

But lo! Madam, who had flown the moment before, returned in hot haste, and flung herself upon that small philanthropist as if he had brought poison; he vanished.

Here was indeed a queer complication! It was a redstart nest without doubt, but who owned the baby? If he were a redstart, why did Mamma refuse help in her hard work, and why did the chestnut-sided insist on helping? If he were a chestnut-sided infant, how did he come in a redstart nest, and what had the redstart to do with him?

These were the problems with which we had to grapple, and we settled ourselves to the work. We placed our seats against neighboring saplings, for backs, and we first critically examined that nest. It was surely a redstart's, though at an unusual height, perhaps twenty-five feet, as we had observed the day before when we had both noted in our books that we saw the male feeding the young. Even had the nest not been so plainly a redstart's, the air of that mother was unmistakable. She owned that nest and that baby, there could not be a doubt, and the dapper little personage with chestnut sides was an interloper.

Nearly two hours we watched every movement of the small actors in this strange drama, and in seeking food they often came within six feet of us on our own level, so that we could not mistake their identity.

The poor little mamma was in deep distress. Although her mate was absent, she resented her neighbor's efforts to help in her work, and dashed at him furiously every time she saw him come. Yet she could not stay on guard, for upon her alone devolved the duty of feeding that nestling! So she rushed frantically hither and thither in her mad redstart fashion, brought her morsel and administered it, and then darted angrily after the enemy, who appeared as often as she did, every time with a tidbit for that pampered youngster.

This double duty seemed almost too much for the redstart. Her feathers were ruffled, her tail opened and shut nervously, and at every interval that she could spare from her breathless exertions she uttered in low tones the redstart song, as though calling on that missing lord of hers.

And where was that much needed personage? Had he been killed in these carefully protected and fenced woods, where no guns or collectors were allowed, and trespass notices were as plentiful as blackberries? Not by shooting we were sure; we should have heard a gun at the house. Had, then, an owl paid a twilight visit, and could a redstart be surprised? Or could, perchance, a squirrel have stolen upon him unaware? We shall never know. There's no morning paper to chronicle the tragedies in the bird world; and it would be too pitiful reading if there were.

The most curious thing about the whole performance was the behavior of the chestnut-sided. His manner was as unruffled as Madam's was excited. The most just and honorable cause in the world could not give more absolute self-possession, more dignified persistence, than was shown by this wonderful atom of a bird. He acknowledged her right to reprove him, for he vanished before her outraged motherhood every time; but the moment the chase ended he fell to collecting food, and by the time his assailant had given her bantling a morsel, he was ready with another.

What could be his motive? Was he a charity-mad personage, such as we sometimes see among bigger folk, determined to benefit his kind, whether they would or no? Had he, perchance, been bereaved of his own younglings, and felt moved to bestow his parental care upon somebody? Did he wish to experiment with some theory of his own on another's baby? Was it his aim to coax that young redstart to desert his family and follow after the traditions of the chestnut-sided?

Alas! how easy to ask; how hard to answer!

By this time I had become as absorbed in the drama as my companion. We forgot, or postponed, the blue, and gave the day to study of this case of domestic infelicity. Five long hours we sat there (morning and afternoon) before the stage on which the interesting but agitating play went on; and after tea, just before dark, we came out again. All this time the war between the two still raged, with no abatement of spirit.

Breakfast was not loitered over on the following morning, and we hurried out to our post. The situation was changed a little. The youngster had made up his mind to go out into the world. He had moved as far as the branch, a few inches from the nest, and was still fed on both sides by his zealous providers. Mamma, however, though every time repelling her unwelcome assistant, was not so nervous. Perhaps she realized that a few hours more would end the trouble. She fed, she encouraged, and pretty soon, while we looked, the infant flew to the nearest tree.

Now the chestnut-sided began to have difficulty in following up his self-imposed charge. He took to coming close upon the mother's heels to see where she went.

But this course was attended with the difficulty that the instant she had fed she was ready to turn upon him, which she never failed to do.

After several short flights about the tree, the young bird, grown bolder, perhaps by over-feeding, for surely never nestling was stuffed as that one was, attempted a more ambitious flight, failed, and came fluttering to the ground, much to the dismay of his mamma, who followed him closely all the way.

This was our opportunity, the moment we had waited for; we must see that disputed baby!

My comrade dropped everything and ran to the spot. A moment's scrambling about on the ground, a few careful "grabs" among the dead leaves, and she held the exhausted little fellow in her hand. He was not frightened; but his mother was greatly disturbed at first. We were too interested in this case to heed her, and indeed after a moment's demonstration she flew away and left him in our hands.

We examined him minutely, and I took note of his markings on the spot. There was no doubt about his being a redstart baby, as I had been convinced from the first. When we had settled this, the little one was placed on a branch, where he remained quite calmly, and we left him to his two attendants.

The next morning we found the mother still hard at work in the same part of the woods (we knew her by some feathers she had lost from her breast), but the gallant little warbler was nowhere to be seen.

XX. A CLEVER CUCKOO

"Hark, the cuckoo, weatherwise,
Still hiding, farther onward woos you."

The mysterious bird, around whose name cluster some strange facts as well as absurd fancies; shy and intolerant of the human race, yet bold in protecting his treasures; devoted and tender in his family relations, yet often known in the neighborhood where he passes his days as a mere "wandering voice,"—

"No bird, but an invisible thing,
A voice, a mystery,"—

this bird, the cuckoo, was a stranger to me till one happy day last June, when I came upon him where he could not escape, beside his own nest.

In returning from our daily visit to the woods that morning, my fellow-student turned down a narrow footway connecting the woods with the home-fields, and I followed. She had passed through half its length, her dog close behind her, when our eyes, ever searching the trees and bushes, fell almost at the same instant upon a nest, with the sitting bird at home. It was so near me that I could have touched it, being not more than two feet from the ground, and hardly farther from the path.

Fearing to startle the little mother, whose frightened eyes were fixed upon us, we announced our mutual discovery by a single movement of the hand, and walked quietly past without pausing. Not until we reached the open fields at the end did my comrade whisper, "a cuckoo," and our hearts, if not our lips, sang with Wordsworth, "Thrice welcome, darling of the spring," for the nest of this shy bird we hardly dared hope to see.

After the morning of our happy discovery the cuckoo path became part of our regular route home from the woods. Our first care was to dispel the fears of the bird, and accustom her to seeing us, so for several days we passed her without pausing, though we looked at her and spoke to her in low tones as we went by.

Three times she flew at sight of us, but on the fourth morning she remained, though with tail straight up and ready for instant flight. But finding that we did not disturb her, she calmed down, and became so fearless that she did not move nor appear agitated when at last we did stop before her door, spoke to her, and identified her as the black-billed cuckoo.

On the eighth day of our visits it happened that I went to the woods alone. I found the bird at home, as usual, and armed with an opera-glass, I placed myself at some distance to watch her. Half an hour passed before she stirred a feather, but I was not lonely. A mourning-warbler came about, eating and singing alternately, after the manner of his kind, and the pretty trill of the black-throated green warbler came out of the woods. Then a crow mamma created a diversion by helping herself to an egg for her baby's breakfast, when a robin and a vireo —curious pair!—took after her with loud cries of indignation and reproach.

When this excitement was over, the trio had disappeared in the woods, and silence had fallen upon us again, I heard the cuckoo call at a little distance, and in a moment the bird himself alighted on a twig three feet above the nest. He was a beauty, but he appeared greatly excited. He threw up his tail till it pointed to the sky over his head, then let it slowly drop to the horizontal position. This he did three times, while he looked down upon his household, so absorbed that he did not see me at all.

Then the patient sitter vacated her post, and he flew down to the nest. The top was hidden by leaves, so that I cannot positively affirm that he sat on the eggs, but it is certain that he remained perfectly silent and motionless there for forty-five minutes. Then I caught sight of Madam returning. She came in from the woods, behind and at the level of the nest; there was a moment's flutter of wings, and I saw that her mate was gone, and she in her usual place.

The next day there was a change in the programme. It happened that I arrived when the mother was away, and the head of the household in charge. No sooner did I appear on the path than he flew off the nest with great hustle, thus betraying himself at once; but he did not desert his post of protector. He perched on a branch somewhat higher than my head, and five or six feet away, and began calling, a low "coo-oo." With every cry he opened his mouth very wide, as though to shriek at the top of his voice, and the low cry that came out was so ludicrously inadequate to his apparent effort that it was very droll. In this performance he made fine display of the inside of his mouth and throat, which looked, from where I stood, like black satin.

The calls he made while I watched him sounded so far off that if I had not been within six feet of him, and seen him make them, I should never have suspected him:

"A cry
Which made me look a thousand ways,
In bush and tree and sky."

Finding that his voice did not drive me away, the bird resorted to another method; he tried intimidation. First he threw himself into a most curious attitude, humping his shoulders and opening his tail like a fan, then spreading his wings and resting the upper end of them on his tail, which made at the back a sort of scoop effect. Every time he uttered the cry he lifted wings and tail together, and let them fall slowly back to their natural position. It was the queerest bird performance I ever saw.

During all this excitement there sounded from a little distance a low single "coo," which, I suppose, was the voice of his mate. Not wishing to make a serious disturbance in the family, and seeing that he was not to be conciliated, I walked slowly on, looking in the nest as I passed. It contained one egg that looked like a robin's, and beside it a small bundle of what resembled black flesh stuck full of white pins. This, then, was the cuckoo baby; surely an odd one!

On the third day after this experience we were fortunate enough again to find the nest uncovered. A second youngster lay beside the first, and the two entirely filled the nest. They were perhaps two and a half inches long, and resembled, as said above, mere lumps of flesh. After looking at the young family, we seated ourselves a little way off to wait for some one to come home.The place the cuckoo had chosen to nest was one of the most attractive spots on the grounds, an opening in the woods in which, after the loss of the trees, had grown up a thicket of wild berries. The bushes were nearly as high as one's head, and so luxuriant that they made an impenetrable tangle, through which paths were cut in all directions, and kept open by much work each year.

In the middle of the opening was a clump of larger saplings, around the foot of two or three very tall old basswood-trees, part of the original forest. It was the paradise of small fruits. Early in the season elderberries beripened, and offered food to whoever would come. Before they were gone the bushes were red with the raspberry, and blackberries were ready to follow; choke-cherries completed the list, and lasted till into the fall. The insect enemies of fruit were there in armies.

Its constant supply of food, its shelter from the winds on every side, and its admirable hiding-places for nests, made this warm, sunny corner the chosen home of many birds. Warblers were there in early spring, heard, though not always seen. Veeries nested on its borders, woodpeckers haunted the dead trees at the edge, and all the birds of the neighborhood paid visits to it.

We had not waited long when the head of the cuckoo family appeared. He saw us instantly, and, I regret to say, was no more reconciled to our presence than he had been on the previous occasion; but he showed his displeasure in a different way. He rushed about in the trees, crying, "cuck-a-ruck, cuck-a-ruck," running out even to the tip of slender branches that seemed too slight to bear his weight. When his feelings entirely overcame him he flew away, and though we remained fifteen minutes, no one came to the nest.

The day after this display of unkindly feeling toward us we passed down the cuckoo path, saw Madam on the nest, and at once determined to wait and see what new demonstration her mate would invent to express his emotions. My comrade threw herself down full length on the dead leaves beside the path, where she could bask in the sunlight, while I sat in the shade close by.

After some time we saw the cuckoo stealing in by a roundabout back way through the low growth in the edge of the wood. He was coming with supplies, for a worm dangled from his beak. He had nearly reached the nest—in fact was not two feet away—when his eyes fell upon us. He stopped as if paralyzed. We remained motionless, almost breathless, but he did not take his eyes off us, nor attempt to relieve himself of that worm. Still we did not move; arms began to ache, feet tingled with "going to sleep," every joint stiffened, and I began to be afraid I should find myself turned to stone. Still that bird never moved an eyelid, so far as we could see.

It was fully twenty-five minutes that we three stared at each other, all struck dumb. But Nature asserted herself in us before it did in him. The sun was hot, and the mosquitoes far from dumb. We yielded as gracefully as we could under the circumstances, and left him there as motionless as a "mounted specimen" in a glass case.

The next morning we started out rather earlier than usual, half expecting to find Master Cuckoo grown to that perch. It appeared, however, that he had torn himself away, for he was not to be seen. The little mother, who was on the nest, had readily learned that we intended no harm, but her peppery little spouse learned nothing; he was just as unreconciled to us the last day as the first.

This time he tried to keep out of sight. First we heard his call far off, then a low "cuck-a-ruck" quite near, to which she replied with a gentle "coo-oo" hardly above her breath.

It was soothing, but it did not altogether soothe. He came up from behind us with another dangling worm in his mouth, slipped silently through the bushes to the nest, and in a moment departed by the back way without a word. Then we went nearer, looked once more upon the shy but brave little mother, and went our way.

We did not suspect it, but that was our last sight of the cuckoo family at home; the next day the place was empty and deserted.

I was smitten with remorse. Were we the cause of the calamity? Had the poor birds carried off the babies? Or had, perchance, another nest tragedy occurred? We looked carefully; there were no signs of a struggle. They had apparently flown in peace. Yet six days before one was still in the egg and the other newly hatched. Only two days ago the pair looked like tiny black cushions covered with white pins, and not a quarter the size of the parents. Moreover, they had been sat upon every day.

In this painful uncertainty we were obliged to leave the matter; but although we saw no more of them, they did not pass out of our minds. Every day we looked in the woods and listened for cuckoo voices, but every day we were disappointed, until about eleven days later.

We were walking slowly down what we called the veery road in the woods, far over the other side from the cuckoo's nest, when we heard a very low but strange baby cry in some thick bushes. It was a constant repetition of one note, a gentle "tut, tut, tut."

We were naturally eager to see the youngster, and we carefully approached the spot. As we came near, a cuckoo flew up, scrambled through a tree, and disappeared. Could it be a cuckoo baby we had heard? In an instant the fugitive seemed to think better of her intention to fly. Perhaps she was conscience-smitten for deserting the little one, for she returned in plain sight, though at some distance. She began at once calling and posturing, clearly for our benefit. We, of course, understood her tactics. She wished to draw us away from the neighborhood of her infant, and as it was impossible to penetrate the thicket, and we did not enjoy torturing an anxious mother, we decided to yield to her wishes, and see what she would do.

She cried every moment, "tut, tut, tut," in a low tone, and ten or twelve times repeated. At the same time she lifted her long tail, and slowly let it fall, with a beautiful and graceful motion. She crouched on the branch, and put her head down to it, then suddenly rose and threw up her head and tail, making herself as conspicuous and as remarkable as she could. We moved a little toward her. That encouraged her to go on; and easily, in a sort of careless, inconsequent way, she hopped to the next branch farther. So we let ourselves be drawn away, she keeping up all the time the low call, while the infant, which we are sure was there, had become utterly silent.

She was a beautiful creature, a picture of grace; and when she had beguiled us some distance away from where we heard the baby-cry, she suddenly slipped behind a branch and was gone; and we felt repaid for missing the young one by the beautiful exhibition she had made of herself. We never saw her again.

XXI. TWO LITTLE DRUMMERS

Last summer I made the acquaintance of an outlaw; an unfortunate fellow-creature under the ban of condemnation, burdened with an opprobrious name, and by general consent given over to the tender mercies of any vagabond who chooses to torture him or take his life. One would naturally sympathize with the "under dog," but when, instead of one of his peers as opponent, a poor little fellow, eight inches long, has arrayed against him the whole human race, with all its devices for catching and killing, his chances for life and the pursuit of happiness are so small that any lover of justice must be roused to his defense, if defense be possible.

The individual of whom I speak is, properly, the yellow-bellied woodpecker, though he is more commonly known as the sapsucker, in some places the squealing sapsucker; and I hailed with joy his presence in a certain protected bit of woods, a little paradise for birds and bird lovers, where, if anywhere, he could be studied. There is some propriety in applying to him the strange epithet "squealing," I must allow, for the bird has a peculiar voice, nasal enough for the conventional Brother Jonathan; but "sapsucker" is, in the opinion of many who have studied his ways, undeserved. Dr. Merriam, even while admitting that the birds do taste the sap, says positively, "It is my firm belief that their chief object in making these holes is to secure the insects which gather about them."

My introduction to the subject of my study took place just after sundown on a beautiful June evening. We were riding up from the railway station, three miles away. The horses had climbed to the top of the last hill, and trotted gayly through a belt of fragrant woods which reached like an arm around from the forest behind, as if lovingly inclosing the attractive scene—a pleasant, old-fashioned homestead, with ample lawn sloping down toward the valley we had left, and looking away over low hills to the unbroken forests of the Adirondacks.

At this moment there arose a loud, strange cry, of distress it seemed, and I turned hastily to see a black and white bird, with bright red crown and throat, bounding straight up the trunk of an elm-tree, throwing back his head at every jerk with a comical suggestion of Jack's "Hitchety! hatchety! up I go!" as he joyously mounted his beanstalk, in the old nursery story. There was surely nothing amiss with this little fellow, and, knowing almost nothing of the

"Greys, whites, and reds,
Of pranked woodpeckers that ne'er gossip out,
But always tap at doors and gad about,"

I eagerly demanded his name, and was delighted to hear in answer, "The sapsucker." I was delighted because I hoped to see for myself whether the bird merited the offensive name bestowed upon him, or was the victim of hasty generalization from careless observation or insufficient data, like others of his race. The close investigations of scientific men have reversed many popular decisions. They have proved the crow to be the farmer's friend, most of the hawks and owls to be laborers in his interest, the kingbird to fare almost entirely upon destructive insects rather than bees, and other birds to be more sinned against than sinning.

The first thing noted was the sapsucker's peculiar food-seeking habit. One bird made the lawn a daily haunt, and we, living chiefly on the veranda, saw him before us at all hours, from dawn to dusk, and thus had the best possible chance to catch him in mischief, if to mischief he inclined. He generally made his appearance flying in bounding, wave-like fashion, uttering his loud mournful cry, which, though an apparent wail, was evidently not inspired by sadness. Alighting near the foot of a tree-trunk, with many repetitions of his complaining note, he gayly bobbed his way up the bark highway as if it were a ladder. When he reached the branches, he flew to another tree.

This bird's custom of delivering his striking call as he approached and mounted a tree not far from his "food tree" may be a newly acquired habit; for Dr. Merriam, who observed this species ten years ago on the same place, says that he "never heard a note of any description from them, either while in the neighborhood of these trees, or in flying to and fro between them and the forests." On his own trees the sapsucker was not in such haste, but lingered about the prepared rings, evidently taking his pick of the insects attracted there.

The array of traps prepared for the woodpecker's use was most curious, and readily explained how he came by his name. The clever little workman had selected for his purpose two trees. One was a large elm, and around its trunk, about fifteen feet from the ground, he had laboriously cut with his sharp beak several rings of cups. These receptacles were somewhat less than half an inch in diameter, and nearly their own width apart, and the rings encircled the trunk as regularly as though laid out with mechanical instruments. His second depot of supplies was one of a close group of mountain ashes, which seemed to spring from one root, and were thickly shaded by leaves to the ground.

The elm would naturally attract the high-flying insects, and the ash those which stay nearer the earth, though I do not presume to say that was the bird's intention in so arranging them. The mountain-ash trunk was perforated in a different way from the elm, the holes being in lines up and down, and the whole trunk covered five or six feet above the root. These places were not at all moist or sticky on the several occasions when I examined them, and both trees were in a flourishing condition.

The habit of the author of this elaborate arrangement was to fly from one tree to the other almost constantly. It appeared to lookers-on that he visited the traps on one and secured whatever was caught or lingered there, then went to the other for the same purpose; thus allowing insects a chance to settle on each while he was absent. At almost any hour of the day he could be found vigorously carrying on his insect hunt in this singular fashion.

It was too late in the season to see the sapsucker in his most frolicsome humor, although occasionally we met in the woods two of them in a lively mood, eagerly discussing in garrulous tones their own private affairs, or chasing each other with droll, taunting cries, some of which resembled the boy's yell, "oy-ee," but others defied description. During courtship, observes Dr. Merriam, they are inexpressibly comical, with queer rollicking ways and eccentric pranks, making the woods ring with their extraordinary voices. At this time, early in June, the season of woodpecker wooing was past. Each little couple had built a castle in the air, and set up a household of its own, somewhere in the woods surrounding the house.

The two storehouses on the lawn seemed to belong to one family, whose labor alone had prepared them; certainly they were the property of the sapsuckers. But the bird world, like the human, has its spoilers. A frequent visitor to the elm, on poaching bent, was a humming-bird, who treated the beguiling cups like so many flowers, hovering lightly before them, and testing one after another in regular order. The owner naturally objected, and if present flew at the dainty robber; but the elusive birdling simply moved to another place, not in the least awed by his comparatively clumsy assailant. Large flies, perhaps bees also, buzzed around the tempting bait, and doubtless many paid with their lives for their folly.

The most unexpected plunderer of the sapsucker stores was a gray squirrel, who lay spread out flat against the trunk as though glued there, body, arms, legs, and even tail, with head down and closely pressed against the bark. I cannot positively affirm that he was sucking the sap or feeding upon the insects attracted to it, but it is a fact that his mouth rested exactly over one of the rings of holes; and his position seemed very satisfactory, for some reason, for he hung there motionless so long that I began to fear he was dead. All these petty pilferers may possibly have regarded the treasure as nature's own provision, like the flowers, but one visitor to his neighbor's magazine certainly knew better. This was the brilliant cousin of the sapsucker, the red-headed woodpecker, whose vagaries I shall speak of a little later.

Nothing about the tri-colored family is more interesting than its habit of drumming—

"The ceaseless rap
Of the yellow-hammer's tap,
Tip-tap, tip-tap, tip-tap-tip.
'Tis the merry pitter-patter
Of the yellow-hammer's tap."

Whether or not it is mere play is perhaps yet an open question. The drumming of the sapsucker, one of the most common sounds of the woods and lawn, seemed sometimes simply for amusement, but again it appeared exceedingly like a signal. A bird frequently settled himself in plain sight of us, on one of the trespass notices in the woods, and spent several minutes in that occupation, changing his place now and then, and thus producing different sounds, whether with that intention or not.

Now he would tap on top of the board, again down one side, and then on a corner, but always on the edge. Nor was it a regular and monotonous rapping; it was curiously varied. One performance that I carefully noted down at the moment reminded me of the click of a telegraph instrument. It was "rat-tat-tat-t-t-t-t-rat-tat," —the first three notes rather quick and sharp, the next four very rapid, and the last two quite slow. After tapping, the bird always seemed to listen. Often while I was watching one at his hammering, a signal of the same sort would come from a distance. Sometimes my bird replied; sometimes he instantly flew in the direction from which it came. Around the house the woodpeckers selected particular spots to use as drums, generally a bit of tin on a roof, or an eave-gutter of the same metal. A favorite place was the hindquarters of a gorgeous gilded deer that swung with the wind on the roof of the barn.

So closely were they watched that the sapsuckers themselves were like old acquaintances before the babes in the woods began to make themselves heard aloud. No sooner had these little folk found their voices than they made the woods fairly echo. Cry-babies in feathers I thought I knew before, but the young woodpecker outdoes anything in my experience. It is no wonder that the woodpecker mamma sets up her nursery out of the reach of prowlers of all sorts; so loud and so persistent are the demands of her nestlings that they would not be safe an hour, if they could be got at. The tone, too, must always arrest attention, for it is of the nasal quality I have mentioned. The first baby whisper, hardly heard at the foot of the tree, has a squeaky twang, which strengthens with the infant's strength, and the grown-up murmurs of love and screams of war are of the same order.

It was during the nest-feeding days that we discovered most of the sapsucker homesteads; for, having many nests nearer our own level to study, we never sought them, and noticed them only when the baby voices attracted our attention. The home that apparently belonged to our bird of the lawn was beautifully placed in a beech-tree heavy with foliage. At first we thought the owner an eccentric personage, who had violated all sapsucker traditions by building in a living tree; but, on looking closely, it was evident that the top of the tree had been blown off, and from that break the trunk was dead two or three feet down. In that part was the opening, and the foliage that nearly hid it grew on the large branches below. Most of the nests, however, were in the customary dead trunks, on which we could gently rap, and bring out whoever was at home to answer our call.

Young woodpeckers are somewhat precocious; or, to speak more correctly, they stay in the nest till almost mature. We see in this family no half-fledged youngster wandering aimlessly about, unable to fly or to help itself, a sight very common among the feathered folk whose homes are nearer the ground. One morning, a young bird, not yet familiar with the mysteries of the world about him, flew into the open window of a room in the house, and for an hour we had a fine opportunity to study him near at hand.

The moment he entered he went to the cornice, and although he flew around freely, he did not descend even so low as the top of the window, wide open for his benefit. He was not in the least afraid or embarrassed by his staring audience, nor did he beat himself against the wall and the furniture, as would many birds in his position; in fact, he showed unusual self-possession and self-reliance.

He was exceedingly curious about his surroundings: tapped the wall, tested the top of picture frames, drummed on the curtain cornice, and closely examined the ceiling. He was beautifully dressed in soft gray all mottled and spotted and barred with white, but he had not as yet put on the red cap of his fathers. While we watched him, he heard outside a sapsucker cry, to which he listened eagerly; then he drummed quite vigorously on the cornice, as if in reply. It was not till he must have been very hungry that he blundered out of the window, as he had doubtless blundered in.

The beauty of the drumming family, at least in that part of the country, is the red-headed woodpecker, which it happened I did not know. The first time I saw one, he was out for an airing with his mate, one lovely evening in June. The pair were scrambling about, as if in play, on the trunk of a tall maple-tree across the lane. They did not welcome our visit, nor our perhaps rather rude way of gazing at them; for one flew away, and the other perched on the topmost dead branch of a tree a little farther off, and proceeded to express his mind by a scolding "kr-r-r," accompanied by violent bows toward us. Finding his demonstration unavailing, he soon followed his mate, and weeks passed before we saw him again, although we often walked down the lane with the hope of doing so.

One beautiful morning, after the hay had been cut from the meadow, and all the hidden nests we had looked at and longed for while grass was growing, were opened to us, I had taken my comfortable folding-chair to a specially delightful nook between a clump of evergreens, which screened it from the house, and a row of maples, elms, and other trees, much frequented by birds. Close before me was a beautiful hawthorn-tree, in which a pair of kingbirds had long ago built their nest.

On one side I could look over to an impenetrable, somewhat swampy thicket, where song sparrows and indigo birds nested; on the other, past the picturesque old-fashioned arbor, half buried under vines and untrimmed trees, far down the pretty carriage-drive between young elms and flowering shrubs, where the bobolink had raised her brood, and the meadow lark had chanted his vesper hymn for us all through June.

Many winged strangers came to feast on the treasures uncovered by the hay-cutter, and then the shy redhead showed himself on our grounds. To my surprise, he was searching the freshly cut stubble not at all like a woodpecker, but hobbling about most awkwardly, half flying, half hopping, seeking some delectable morsel, which, when found, he carried to the side of a tree-trunk, thrust into a crack, and ate at his leisure. The object I saw him treat in this way was as large as a bee, and he was some time in disposing of it, even after it was anchored in the crack. Then, observing that, although a long way off, I was interested in his doings, he slipped around behind the trunk, and peered at me first from one side, then in an instant from the other.

The next performance with which this bird entertained me was poaching upon his cousin's preserves. Sitting one evening on the veranda, looking over the meadow, I heard his low "kr-r-r," and saw him alight upon the sapsucker's elm. Whether he stumbled upon the feast or went with malice aforethought, he was not slow to appreciate the charms of his position. It may have been the nectar from the tree, or the minute victims of its attractions, I could not tell which, but something pleased him, for he devoted himself to the task of exploring the tiny cups his industrious relative had carved, driving away one of the younger members of the family already in possession.

The young bird could not refuse to go before the big beak and determined manner of the stranger, but he did refuse to stay away; and every time he was ousted he returned to the tree, though he settled on a different place. Before the red-head had shown any signs of exhausting his find, the sapsucker himself appeared, and at once fell upon his bigger cousin with savage cries. Disturbed so rudely from his pleasing occupation, the intruder retired before the attack, though he protested vigorously; and so great was the fascination of the spot, that he returned again and again, every time to go through the same process of being driven away.

The raspberry hedge before my windows was the decoy that gave me my best chance to study the red-headed woodpecker. Day after day, as the berries ripened, I watched the dwellers of wood and meadow drawn to the rich feast, and at last, one morning, to my great joy, I saw the interesting drummer alight on a post overlooking the loaded vines. He plainly felt himself a stranger, and not certain of his reception by the residents of the neighborhood, for he crouched close to the fence, and looked warily about on every side.

He had been there but a few moments when a robin, self-constituted dictator of that particular corner of the premises, came down a few feet from him, as if to inquire his business. The woodpecker acknowledged the courtesy by drawing himself up very straight and bowing. The bow impressed, not to say awed, the native bird. He stood staring blankly, till the new-comer proclaimed his errand by dropping into the bushes, helping himself to a berry, and returning to the fence to dispose of his plunder. This was too much; the outraged redbreast dashed suddenly over the head of the impertinent visitor, almost touching it as he passed.

The woodpecker kept his ground in spite of this demonstration, and I learned how a bird accustomed to rest, and even to work, hanging to the trunk of a tree, would manage to pluck and eat fruit from a bush. He first sidled along the top of the board fence, looking down, till he had selected his berry. Then he half dropped, half flew, into the bushes, and sometimes seized the ripe morsel instantly, without alighting, but generally hung, back down, on a stalk which bent and swayed with his weight, while he deliberately gathered the fruit. He then returned to the fence, laid his prize down, and pecked it apart, making three or four bites of it.

After some practice he learned to swallow a berry whole, though it often required three or four attempts, and seemed almost more than he could manage. When he had accomplished this feat, he sat with his head drawn down into his shoulders, as though he found himself uncomfortably stuffed. Having eaten two or three raspberries, our distinguished visitor always picked another, with which he flew away—doubtless for the babies growing up in some dead tree across the lane.

The little difficulty with the robin was easily settled by the stranger. Somewhat later in that first day, he took his revenge for the insulting dash over him by turning the tables and sweeping over the lofty head of the astonished robin, who ducked ingloriously, in his surprise, and called out, "Tut! tut!" as who should say, "Can such things be?" After that Master Robin undertook a closer surveillance of that highway the fence, and might be seen at all hours perched on the tall gatepost, looking out for callers in brilliant array, or running along its whole length to see that no wily woodpecker was hiding in the bushes. He could not be on guard every moment, for his nursery up under the eaves of the barn was full of clamorous babies, and he was obliged to give some attention to them; but the red-head was not afraid of him, and, finding the fruit to his taste, he soon became a daily guest.

Sometimes the spouse of the gay little fellow came also. She was always greeted by a low-whispered "kr-r-r," and the husky-toned conversation between the two was kept up so long as both were there. Now, too, as the male began to feel at home, I saw more of his odd ways. His attitudes were especially comical. Sometimes he clung to the edge of the top board, his tail pressed against it, his wings drooped and spread a little, exposing his whole back, and thus remained for perhaps ten minutes. Again he flattened himself out on top of a post for a sun bath. He sprawled and spread himself, every feather standing independent of its neighbor, till he looked as if he had been smashed flat, and more like some of the feather monstrosities with which milliners disfigure their hats than a living bird.

Another curious habit of my versatile guest was his fly-catching. It is already notorious that the golden-wing is giving up the profession of woodpecker and becoming a ground bird; it is equally patent to one who observes him that the red-head is learning the trade of fly-catching. Frequently, during the weeks that I had him under observation, I saw him fly up in the air and return to the fence, exactly like the kingbird.

All the time I had been making this pleasing acquaintance I had longed in vain to find the red-head's nest. It was probably in the pasture in which we had first met him, where the somewhat spirited cattle in possession prevented my explorations. I hoped at least to see his young family; but July days passed away, and though the bonny couple spent much time among the raspberries, they always carried off the nestlings' share.

In the very last hours of my stay, after trunks were packed, fate relented, and I spent nearly the whole day studying the "tricks and manners" of a red-headed baby. I had returned from my last morning's walk in the woods, and was seated by my window, thinking half sadly that my summer was ended, when I saw the woodpecker come to the raspberries, gather one, and fly away with it. Instead, however, of heading, as usual, for the woods across the pasture, he alighted on a fence near by. A small dark head rose above the edge of a board, opened a bill, and received the berry in it.

Instantly I turned my glass upon that meek-looking head. So soon as the old bird disappeared the young one came up in sight, and in a few moments flew over to the nearer fence, beside the bushes. Then one of the parents returned, fed him two or three times, apparently to show him that berries grew on bushes, and not in the beak, and then departed with an air that said, "There, my son, are the berries; help yourself!"

Left now to his own devices, the little woodpecker was my study for hours. He was like his parents, except that he was gray where they were red, and the white on the wings was barred off with a dark color which on theirs did not appear. Like young creatures the world over, he at once began to amuse himself, working at a hole in the top of a post, digging into it vehemently, and at last, after violent effort, bringing out a stick nearly as long as himself. This he brandished about as a child flourishes a whip, and presently laid it down, worried it, flung it about, and had a rare frolic with it. Tiring of that, he closely examined the fence, going over it inch by inch, and pecking every mark and stain on it.

When startled by a bird flying over or alighting near him, he sprang back instantly, slipped over behind the fence or post, and hung on by his claws, leaving only his head in sight. He was a true woodpecker in his manners; bowing to strangers who appeared, driving away one of his sapsucker cousins who came about, and keeping up a low cry of "kr-r-r" almost exactly like his parents. He showed also great interest in a party of goldfinches, who seemed to have gone mad that morning.

Finally the thought of berries struck the young red-head. He began to consider going for them. One could fairly see the idea grow in his mind. He leaned over and peered into the bushes; he hitched along the fence, a little nearer, bent over again, then came down on the side of the board, and hung there, with body inclined toward the fruit. After many such feints, he actually did drop to the second board, and a little later secured a berry, which he took to the top of the post to eat. In spite of the fact that he was amply able to help himself, as he proved, he still demanded food when his parents came near, bowing and calling eagerly, but not fluttering his wings, as do most young birds.

Nearly all day the little fellow entertained himself; working industriously on the fence, hammering the posts as if to keep in practice, as children play at their parents' life work, and varying these occupations with occasional excursions into the bushes for berries. The notion of flying away from where he had been left never appeared to enter his head. He seemed to be an unusually well-balanced young person, and intelligent beyond his years—days, I should say.

XXII. FROM MY WINDOW

The best place I have found for spying upon the habits of birds is behind a blind. If one can command a window with outside blinds, looking upon a spot attractive to the feathered world, he will be sure, sooner or later, to see every bird of the vicinity. If he will keep the blinds closed and look only through the opened slats, he will witness more of their unconstrained free ways than can possibly be seen by a person within their sight, though he assume the attitude and the stolidity of a wooden figure. Says our nature-poet, Emerson:

"You often thread the woods in vain
To see what singer piped the strain.
Seek not, and the little eremite
Flies forth and gayly sings in sight."

And the bird student can testify to the truth of the verse.

Many times, after having spent the morning in wandering about in the bird haunts of a neighborhood, I have returned to my room to write up my note-book, and have seen more of birds and bird life in an hour from my window than during the whole morning's stroll.

One of my windows, last summer, looked out upon an ideal bird corner: a bit of grass, uncut till very late, with a group of trees and shrubs at the lower boundary, and an old board fence, half buried in luxuriant wild raspberry bushes, running along one side. It was a neglected spot, the side yard of a farmhouse; and I was careful not to enter it myself so often as to suggest to the birds that they were likely to see people. It had the further advantage of being so near the woods surrounding the house, that the shy forest birds were attracted to it.

No sooner would I seat myself, pen in hand, than chirps and twitters would come from the trees, a bird alight on the fence, or else a red squirrel come out to sun himself. Of course the pen gave way to the opera-glass in a moment, and often not a line of the note-book got itself written till birds and squirrels had gone to bed with the sun.

The group of trees which bounded my view at the end of this outdoor study I called the "locust group." It consisted of a locust or two, surrounded by a small but close growth of lesser trees and shrubs that made a heavy mass of foliage. There were a few young ashes, two or three half-grown maples, a shadberry bush, and wild raspberry vines to carry the varied foliage to the ground. Inside this beautiful tangle of Nature's own arranging, was a perfect tent, so thickly grown near the ground that a person could hardly penetrate it without an axe, but open and roomy above, with branches and twigs enough to accommodate an army of birds.

Behind that waving green curtain of leaves took place many dramas I longed to see; but I knew that my appearance there would be a signal for the whole scene to vanish, and with flit of wings the *dramatis personæ* to make their exit. So I tried to possess my soul in patience, and to content myself with the flashes and glimpses I could catch through an opening here and there in the leafy drapery.

At one corner of the group stood a small dead tree. This was the phœbe's customary perch, and on those bare branches—first or last—every visitor was sure to appear. On the lower branch the robin paused, with worm in mouth, on the way to his two-story nest under the eaves of the barn. On the top spire the warbler baby sat and stared at the world about it, till its anxious parent could coax it to a more secluded perch. From a side branch the veery poured his wonderful song, and the cheery little song sparrow uttered his message of good will for all to hear and heed. It was here the red-headed woodpecker waited, with low "k-r-r-r-r" and many bows to the universe in general, to see if the way were clear for him to go to the fence. Nothing is so good to bring birds into sight as an old fence or a dead tree. On the single leafless branch at the top of an old apple-tree the student will generally see, at one time or another, every bird in an orchard.

This dead tree of the locust group was the regular perch of "the loneliest of its kind," the phœbe, whose big chuckle-head and high shoulders gave him the look of an old man, bent with age. His outline one could never mistake, even though he were but a silhouette against the sky. One of these birds could nearly always be seen on the lowest branch pursuing his business of fly-catcher, and I learned more of the singularly reserved creature than I ever knew before. I found, contrary to my expectation, that he had a great deal to say for himself, aside from the professional performance at the peak of the barn roof which gives him his name.

"Phœbe is all it has to say
In plaintive cadence o'er and o'er,"

sings the poet, but he had not so close acquaintance with him as I enjoyed behind my blind. There were two mud cottages in the neighborhood, and two pairs of birds to occupy them, and no phœbe of spirit will tolerate in silence another of his kind near him. Sparrows of all sorts might come about; juncos and chickadees, thrushes and warblers, might alight on his chosen tree—rarely a word would he say; but let a phœbe appear, and there began at once a war of words. It might be mere friendly talk, but it sounded very much like it was vituperation and "calling names," and I noticed that it ended in a chase and the disappearance of one of them.

Again, whenever a phœbe alighted on the fence he made a low but distinct remark that sounded marvelously like "cheese-it," and several times the mysterious bird treated me to a very singular performance. He hovered like a humming-bird close before a nest, looking into it and uttering a loud strange cry, like the last note of "phœbe" repeated rapidly, as "be-be-be." Was it derision, complaint, or a mere neighborly call? This was not for the benefit of his own family, for he did it before the robin's nest. I thought at first he meant mischief to the young robins, but although he approached very near he did not actually touch them.

The loudest note this bird uttered was, of course, his well-known "phœbe," which he delivered from the peak of the barn (never from the dead tree) with an emphasis that proclaimed to all whom it might concern that he had something on his mind. It was plain that he was a person of cares; indeed, his whole bearing was that of one with no nonsense about him, with serious duties to perform. I wonder if these birds are ever playful! Even the babies are dignified and self-contained. Phœbes in a frolic would be a rare sight. Of the two nests whose owners I had to study, one was on a low beam in the cowbarn, where a person might look in; the other under the eaves of a farm-building close by.

The special policeman of the group and its environs was a robin, who lived in a two-story nest under the eaves of the hay-barn. This bird, after the manner of his family, constituted himself regulator and dictator. He lived in peace with the ordinary residents, but took it upon himself to see that no stranger showed his head near the spot. He chased the crow blackbird who happened to fly over on business of his own, and by calls for help brought the whole robin population about the ears of the intruder.

He also headed the mob of redbreasts that descended one morning upon a meek-looking half-grown kitten, who chanced to cast its innocent eyes upon a robin baby under the trees on another side of the house. The youngster could fly with ease, but he preferred to stay on the ground, for he quickly returned there when I put him on a low branch; and when a robin makes up his mind, arguments are useless. The same robin bullied the red-headed woodpecker, and flew at the kingbird when he brought his young family up to taste the raspberries.

One visitor there was, however, to the fence and the locusts whom Master Robin did not molest. When a prolonged, incisive "pu-eep" in the martial and inspiring tone of the great-crested fly-catcher broke the silence, I observed that the robin always had plenty of his own business to attend to. I admire this beautiful bird, perhaps because he is the inveterate enemy of the house sparrow, and almost the only one who actually keeps that little bully in his proper place. There is to me something pleasing in the bearing of the great-crest, who, though of few inches, carries himself in a manner worthy of an eagle. Even the play of a pair of them on the tops of the tallest dead trees in the woods, though merry enough with loud joyful cries, has a certain dignity and circumspection about it uncommon in so small a bird.

A pair of great-crests were frequent visitors to the fence, where they were usually very quiet. But one day as the male flew over from the woods, his call was answered by a loud-voiced canary, whose cage hung all summer outside the kitchen door. The stranger alighted on a tree, apparently astonished to be challenged, but he replied at once. The canary, who was out of sight on the other side of the cottage, answered, and the droll conversation was kept up for some time; the woods bird turning his head this way and that, eager to see his social neighbor, but unable, of course, to do so.

A little later in the season, when baby birds began to fly about, the locust group became even more attractive. Its nearness to the woods, as already mentioned, made it convenient for forest birds, and its seclusion and supply of food were charms they could not resist. First of the fledglings to appear were a family of crow blackbirds, four of them with their parents. These are the least interesting feathered young people I know, but the parents are among the most devoted. They keep their little flock together, and work hard to fill their mouths. The low cry is husky, but insistent, and they flutter their wings with great energy, holding them out level with the back.

After berries began to ripen, the woodpeckers came to call on us. In my walk in the woods in the morning, I frequently brought home a branch of elder with two or three clusters of berries, which I hung in the small dead tree. In that way I drew some of the woods birds about.

The downy woodpecker was one of my first callers. He came with a sharp "chit-it-it," hung upon the clusters, occasionally head down, and picked and ate as long as he liked. The vigilant robin would sometimes fly at him, and he would leave; but in a moment back he came, and went on with his repast. When the care of an infant fell to him, he brought his charge to the source of supplies.

A farm wagon happened to stand near the dead tree, and on this the young woodpecker alighted, and stood humped up and quiet while his parent went to the berries, picked several for himself, and then proceeded to feed him. This young bird was very circumspect in his behavior. He did not flutter or cry, in the usual bird-baby manner, but received his food with perfect composure. Berries, however, seemed to be new to him, and he did not appear to relish them, for after tasting two or three he flew away. In spite of this he came again the next day, and then he flew over to a cluster himself, and hung, back down, while he ate. He was charming with his sweet low chatter, and very lovely in plumage, white as snow, with dark markings clear and soft.

One of the prettiest of our guests was a young chestnut-sided warbler. He looked much bigger than his papa, as warbler babies often do; but that is probably because the young bird is not accustomed to his suit of feathers, and does not know how to manage them. Some of them appear like a child in his grandfather's coat. The chestnut-sided warbler was himself an attractive little fellow, with a generous desire to help in the world's work pleasant to see in bird or man.

After becoming greatly interested in one we had seen in the woods, who insisted on helping a widowed redstart feed her youngster, and had almost to fight the little dame to do so, we found another chestnut-sided warbler engaged in helping his fellows. Whether it were the same bird we could not tell; we certainly discovered him in the same corner of the woods. This little fellow was absorbed in the care of an infant more than twice as big as himself. "A cowbird baby!" will exclaim every one who knows the habit, shameful from our point of view, of the cowbird, to impose her infants on her neighbors to hatch and bring up. But this baby, unfortunately for the "wisdom of the wise," did not resemble the cowbird family.

We saw the strange pair several times in the woods, and then one day, as I sat at my window trying to write, I heard a new cry, and saw a strange bird fly to the fence. He was very restless, ran along the top board, then flew to another fence, scrambled along a few feet, raising and lowering his tail, and all the time uttering a husky two-note baby-cry. While I was struggling to keep him in the field of my glass long enough to note his points, he went to the dead tree, when the philosophical phœbe sitting there took his case in hand, and made a dash for him. The stranger flew straight over the house, with his assailant in close chase. But in a moment I heard the baby-cry in a maple beside the cottage, while the phœbe calmly returned to his post and gave his mind again to his fly-catching. The young bird was not in range from the window, but when, a few seconds later, I heard the feeding-cry, I could no longer resist the desire to see him.

I forgot my caution, and rushed out of the house, for I suspected that this uneasy visitor was the chestnut-sided's adopted charge. So I found it. There stood the infant, big and clumsy by comparison, calling, calling, forever calling; and stretching up on tiptoe, as it were, to reach him was the poor little warbler, trying to stop his mouth by stuffing him. The foster-parent lingered as if he were weary, and his plumage looked as if he had not dressed it for a week. But the insatiate beggar gave him no peace; with the swallowing of the last morsel began his cry for more. Again, standing within ten feet of him, I noticed the young bird's points, and again I was convinced that he was not a cowbird baby.

The curious antics of a solemn kingbird, who did not suspect his hidden observer, were droll to look upon. He seemed to be alone on the fence, though some silent spectator may have been hidden behind the leaves. He mounted suddenly straight up in the air, with cries, twenty feet or more, then soared down with a beautiful display of his plumage. This he did many times in succession, with an indescribably conscious air, and at last he dropped behind some tall grass in the pasture. It looked exceedingly like "showing off," and who could imagine a kingbird in that role!

But all flourishes were over when, somewhat later, he brought his lovely little family of three to the fence to be treated to berries. It was interesting to see a fly-catcher take his fruit "on the wing," as it were; that is, fly at it, seize it, and jerk it off without alighting. The phœbe picked berries in the same way, when he occasionally condescended to investigate the attraction that brought so many strangers into his quiet corner.

The young kingbirds were sweet and chatty among themselves, and they decidedly approved the berries; but they never lost sight of each other, and kept close together, the little company of three, as I have seen other kingbirds do. One day they came in the rain, feathers all in locks, showing the dark color next the skin, and looking like beggars in "rags and tags," but they were as cheerful and as clannish as ever.

To the locust group, too, came the red-headed woodpeckers; at first the parents, who talked to each other in whispered "kr-r-r-r's," and carried off many a sweet morsel to their family in the woods; later, one youngster, who took possession of the fence with the calm assurance of his race, and when I left the place had apparently established himself there for the season.

Many others alighted on the fence; the junco, with his pretty brown bantling and his charming little trilling song; the crow baby, with its funny ways and queer cry of "ma-a-a;" the redstart, who

"Folds and unfolds his twinkling tail in sport;"

the flicker mamma, with her "merry pitter-patter" and her baby as big as herself. Even the sapsucker from the lawn had somehow heard the news that a feast was spread near the locusts, and came over to see.

Birds were not the only frequenters of the fence and the berry bushes. There were squirrels, gray and red, and chipmunks, who sat up pertly on a post, with two little paws laid upon their heart in theatrical attitude, as who should say, "Be still, my heart," while they looked the country over to see if any lurking member of the human family were about. The red squirrels were the most amusing, for they were very frolicsome, indulging in mad chases over and under the fence, through the trees, around the trunks, so rapidly that they resembled a red streak more than little beasts.

One squirrel adopted the fence as his regular highway, and the high post of the farm gate as his watch-tower. He often sunned himself, lying on his face, with his legs and his tail spread out as flat as if he had been smashed. His presence scared the birds from the neighborhood, and I undertook to discourage him. I went out one day when I saw him near the fence. The squirrel made up his mind to pass over the gate and get into the locust, but I posted myself quite near, and he did not like to pass me. Giving up his plan is no part of a squirrel's intention, however, and every moment he would scramble up a few feet one side of me, with the design of running past me. As soon as his sharp black eyes showed above the top board I cried "Shoo!" He understood my motion, and doubtless would if I had said "Scat!" or "Get out!" (What should one say to a squirrel?)

He dashed behind his barricade and disappeared. But he did not "stay put;" in two seconds he tried it again, and again his discouraging reception drove him back. He grew wary, however, and pretty soon I began to notice that every time he made his dash to the top he was a few inches nearer the gate, which stretched like a bridge from the fence to the locust-tree, and of course so much nearer me. At last, advancing thus inch by inch, he came up close to the gate, so near I could have put my hand on him—that is, I could have put my hand on the place he occupied, for he did not stay to be caressed; he flew across the gate, sprang three or four feet into the tree, and was out of sight before I could lift a finger. This passage having been successfully made, he felt that he was safe, and could afford to be saucy. He began the usual scold. Then I tossed a little stick up toward him, as a reminder that human power is not limited by the length of an arm, and he subsided.

Once when he came up to the fence top, before his grand dash, I laughed at him. Strange to say, this made him furious. He reviled me vehemently. No doubt, if I had understood his language, I should have been covered with confusion, for I confess that he could make a very good point against me. What business had I, an interloper in his dominion, to interfere with his rights, or to say whether he should dine off birds or berries?

XXIII. THE COMICAL CROW BABY

Nothing in the world of feathers is so comical as a crow baby, with its awkward bows and ungainly hops, its tottering steps on the fence and its mincing, tight-boot sort of gait on the ground, its eager fluttering when it has hopes of food, and its loud and unintermitting demand for the same.

My window overlooked a long stretch of cattle pastures and meadows still uncut, bounded on one side by woods, and in the middle of this valley unvisited by man, the crows of the neighborhood established a training school for their youngsters. A good glass let me in as unsuspected audience, and I had views of many interesting family scenes, supposed by the wary parents to be visible only to the cows stolidly feeding on the hillside. In this way I had all the fun and none of the trouble of the training business.

It is astonishing how completely the manner of the adult crow is lacking in his young offspring, whose only external difference is the want of a tail. Must we then conclude that the dignity of a bird depends upon the length of his tail? We are accustomed to regard the crow as a grave and solemn personage with a serious role in life; and indeed life is such a constant warfare to him that I cannot see how he finds any enjoyment in it. Lowell says of him at one period:

"The crow is very comical as a lover, and to hear him try to soften his croak to the proper Saint Preux standard has something the effect of a Mississippi boatman quoting Tennyson."

If he is droll as a lover, he is much more entertaining as an infant. The first I knew of the new use of the pasture, I heard one morning a strange cry. It was loud and persistent, and sounded marvelously like "Ma-a! Ma-a!" Mingled with it I heard the vigorous cries of crows.

I looked over into the pasture, and there I first saw the crow baby, nearly as big and black as his mamma, but with no tail to speak of. He sat—not stood—on the rail fence, bawling at the top of his hoarse baby-voice, "Ma! Ma! Ma!" and as he grew impatient he uttered it faster and faster and louder and louder, drawing in his breath between the cries, and making it more like "Wah! Wah!" Whenever mamma flew over he followed her movement with his eyes, turning his head, and showing an eager, almost painful interest, till some one took pity on him and fed him. As he saw food approaching his voice ran up several tones higher, in laughable imitation of a human baby cry. This note is of course the promise of a "caw," but the *a* is flattened to the sound of *a* in bar, which makes it a ludicrous caricature of our own first utterances.

But sometimes mamma did not heed the cries, and sailed calmly by, alighting a few rails beyond her hungry infant, though he held out his fluttering wings in the bird-baby's begging way, exactly as does a young warbler who wouldn't be a mouthful for him. Then the little fellow would start up on unsteady legs, to walk the rail to reach her, balancing himself with outspread wings, and when he got beside her, put his beak to hers in a coaxing way that I don't see how any mother could resist. But this wise dame had evidently hardened her heart. She probably wanted him to learn to help himself, for she dropped to the ground, and went wading about in the wet grass and mud, and at length flew off without giving him a morsel. Then the disappointed youngster cuddled up to a brother crow baby, and both lifted up their voices and lamented the emptiness of the cold, cold world.

Perhaps the most comical performance of this clumsy baby was his way of alighting on a fence when he had been flying. He seized the board with his claws, which clung for dear life, while his body went on as it was going, with the result almost of a somersault. He tried to learn, however. He made great efforts to master the vagaries of fences, the irregularities of the ground, the peculiarities of branches. He persistently walked the rail fence, though he had to spread both wings to keep his balance. Then he climbed to the top of the rail which stood up at the corners, and maintained his position with great effort, but never gave up the attempt.

These interesting young folks dote on fences, after they get used to them, and not having learned to recognize them as devices of the enemy, capable of concealing a trap of some sort, they will come quite near a house when they see no one about. So I, behind my blind, had excellent chance to watch their ways. For I try to keep my window view good by contenting myself with what I can see from it, and never going out to give the birds a notion that they must look out for visitors.

One day when the grass had been cut from the meadow before the house, and I had encamped under the shade of a big maple to see how the kingbirds were coming on in nesting, I noticed a young crow walking in the hot stubble, trying to find something to eat. He wandered about looking in vain to see something attractive. A robin who was also engaged in a food-hunt came and "took his measure," looking sharply at him as if to decide whether it was his duty to go for him. He plainly recognized the youthfulness of the intruder, for after a moment's study he passed on, attending to his own business, while the young crow stared at him in open-mouthed curiosity. At last the crow baby picked up an object—I could not tell what—which hung from his beak while he balanced the probabilities of its being good, aiding his deliberations by a gentle lift of the wings which looked like a shrug of the shoulders.

He decided to risk it, and swallowed, but he instantly choked it up, and for some time shook his head as if to get rid of even just the memory of it. When, a few minutes after this disastrous experience, he heard another baby utter the cries that indicate being fed, it did seem to suggest to him an easier way of getting satisfaction out of life. He spread his wings, flew to a tree and began to call.

To be a crow mamma is no sinecure. My heart went out to the poor souls who must be torn between anxiety for their dear "cantankerous" offspring, and fear of their deadly enemy, man. I watched with deep interest their method of training. One day I saw a baby get an object lesson in his proper attitude toward mankind, in this way. An old and a young crow were nearer the house than usual, and I walked down toward the fence to see why. The instant my head appeared, the elder flew with terrific outcry, for which of course I did not blame the poor creature, since mankind has proved itself her bitterest foe. The infant was nearly frightened to death, and followed as quickly as his awkward wings would carry him. I do not like to figure as "Rawhead and bloody-bones" in the nursery of even a crow baby, so I tried several times to redeem the bad name of my race. But to no avail; that subtle mamma had acquired her wisdom by experience, and she knew me as one of a species quite capable of murdering an innocent crow baby.

I was interested to see the young family in the pasture taking lessons in following, or flying in a flock. There was great excitement and calling, and all flew, excepting one, who stood quietly on a big stone by himself. They simply circled around and alighted again, so it plainly was only an exercise. But the baby who did not learn the lesson and follow, was punished by one of the grown-ups, who flew directly against him on the return, and knocked him off his perch; the hint was taken, and the next time they flew no one stayed behind.

Day by day the excitement in the crow world grew, and new families appeared in the pasture as fast as old ones got out. The rails of the fence were always occupied by young ones—though never more than five or six at a time—crying and shrieking and calling for "Ma-a!" and old ones all the time flying about half distracted, cawing and trying, I suppose, to enforce some order and discipline among the unruly rogues. Order, however, was quite a secondary consideration; the pressing duty of the hour was feeding.

A crow parent on a foraging expedition is a most unwelcome visitor to the farmer with young chickens, or the bird-lover interested in the fate of nestlings. Yet when I saw the persecuted creature in the character of provider for four hungry and ever clamorous mouths, to whose wants she is as alive as we are to the wants of our babies, I took a new view of crow depredations, and could not see why her children should not have a chicken or a bird for breakfast, as well as ours. Poor hunted crow, against whom every man's hand is raised! She feels, with reason, that every human being is a deadly enemy thirsting for her life, that every cylinder pointed upward is loaded with death, that every string is a cruel snare to entangle and maim her—yet whose offspring, dear as ours to us, clamor for food. How should she know that it is wrong to eat chickens; or that robin babies were made to live and grow up, and crow babies to die of starvation?

The farmer ignores the millions of insects she destroys, and shoots her for the one chicken she takes, though she has been amply proved to be one of his most valuable servants. The kingbird and the oriole worry her life out of her because her babies like eggs—as who does not!

In fact, there are, emphatically, two sides to the crow question, and I take the side of the crow.

XXIV. A MIDSUMMER WOOING

The "sweet June days" had passed, and bird nesting was nearly at an end. Woods and fields were bubbling over with young bird notes, and the pretty cradles on tree and shrub were empty and deserted. A few motherly souls, it is true, were still occupied with their second broods, but, in general, feathered families were complete, and the parents were busy training their little folk for life.

One bird, however, the charming, sweet-voiced goldfinch,

"All black and gold, a flame of fire,"

still held aloof, as is his custom. He does not follow the fashion of his fellows; he resists the allurements of the nesting month; he waits. Whether it be for a late-coming insect necessary to the welfare of his nestlings, or for the thistle silk which alone makes fit cushion for his delicate spouse and her "wee babies," opinions differ.

But though goldfinch nests were not set up, goldfinch wooing went on with enthusiasm; the summer air rang with sweetest song, and the graceful wave-like flight charmed us from morning till night. The courtship of the bird of July is a beautiful sight. He is at all times peculiarly joyous, but at this season his little body seems hardly able to contain him; so great is his rapture, indeed, that it infects and inspires the most matter-of-fact student. Our bird-loving poet Celia Thaxter must have seen him in loverly mood when she thus addressed him:

"Where do you hide such a store of delight,
O delicate creature, tiny and slender,
Like a mellow morning sunbeam bright,
Overflowing with music tender?"

At all hours of these enchanted days, whether fair or foul, the winsome little fellows were flying hither and thither, singing and calling in ecstatic tones, bounding through the air, and hardly pausing long enough to eat. July was fast slipping away when the excitement deepened and matters grew more serious. Then the observer, if he were wary, might catch occasional glimpses of puzzling scenes, mysteries of bird life that could not be unraveled because he did not see the whole.

At one time the student came upon a scene like this: Two or three of the little dames in olive and gold hopping about on an evergreen tree, ostensibly eating, calling, in their enticing voices, "sw-e-e-t!" and to all appearance unconscious of the presence of two of their bright young wooers, sitting in perfect silence on an upper branch. Suddenly from this happy party one of the damsels flew, when instantly one of the black-winged suitors flashed out in pursuit. On she went, flying madly, encircled one tree, dashed to another, and around that, passed up and down, here and there, this way and that, but everywhere with her follower close after her, singing at the top of his voice, till they disappeared in the distance.

Can the goldfinch wooing be a sort of Comanche affair? Is the little bride won by force? Or is she, perchance, like some of her sisters of larger growth, who require a "scene" of some sort to make them "name the day"?

Again, attracted by loud eager singing, the student found a pair who were apparently fighting—the peaceful goldfinch! They flew up close together, they almost clinched, then flew away to a group of trees, under, over, around, between, through, and beyond they went, never six inches apart, and he singing furiously all the time. At last, just as the looker-on expected to see them grapple, they calmly alighted on a tree eight or ten feet from each other. Nothing but a frolic, obviously!

Another curious performance of this July wooing was several times noted. Hearing a strange and unfamiliar cry, in a tone of distress, I drew cautiously near, and found, on a low branch, one of the goldfinch maidens, uttering the plaintive notes, which, by the way, were afterwards very common about the nests. She held in her beak something which might be a tiny green worm, or a bit of nesting material, and she called constantly, looking about this way and that, as if seeking some one. After a while a male goldfinch appeared on the next tree, but he did not act in the least as if invited by her call. He seemed merely to be interested as any bird would be by her evident excitement. He watched her calmly, but did not offer to follow when at last she flew.

Time, true to his reputation, was hurrying away even these sweet summer days, and still the love affairs of our little beauties seemed no nearer settlement than at first. In the opinion of impatient observers, their wooing was as long drawn out as that of Augustus and Araminta in an old-fashioned three-volume novel. Their manners, too, ludicrously suggested the behavior of the bigger pair; first he would follow her about, sing to her, parade himself, and show off; then she coquetted, and charmed him with her bewitching and altogether indescribable call, "sw-e-e-t." Then they were off in a whirl of excitement together, flitting hither and thither, singing and dancing through the air, life showing its rosiest hue.

All things come to an end—in time. By the middle of the month the ecstasies of goldfinch youth were toned down, and the presence of dainty nests here and there proved that madam at least had settled to work, making preparation for her long, patient brooding.

The tall grass in the meadow in front of the house was about this time laid low; nodding daisies—white and yellow—plumy meadow-grass and plain timothy, devil's paintbrush and soft purple grass flowers, alike lay in long rows dying on the ground. Delighted at last to possess the places so long tabooed to us by the heavy crop, my comrade and I went out the next morning on discoveries bent. The nook in which we rested after our walk—she on the fresh sweet hay in the broad sunshine, and I in the shade close by—offered a rare combination of seclusion with perfect security. It was within call from the veranda, yet completely hidden from it by a dense clump of evergreens.

We had hardly settled ourselves when we noticed three lively goldfinches frolicking about the top of a tall maple-tree not far off. While we idly speculated about them, wondering if they had no mates, and if the goldfinches were not going to build this year, the eyes of my friend, who was lying on the ground, fell upon the nest. It was near the end of a lower branch of the maple, ten or twelve feet from the ground, and the little dame was at that moment working upon it. She was so deeply absorbed in her occupation that she did not even notice us, and we studied her movements with interest, till the haymakers came with wagon and oxen, and much talking and shouting, to gather up their fragrant loads, which on that side of the field stood in small stack's all ready.

Once again, in spite of long experience, I was amazed to see how deaf and blind are people to what goes on about them. "We see only that which concerns us," says some one, and since the farmer, with whole mind bent upon making a firm and symmetrical load, did not concern himself with bird affairs, goldfinch work went on without hindrance. The half-loaded wagon paused under the chosen branch, where the man could have laid his hand upon the nest, but the small builder went in and out, calling and fluttering around as freely as if he were not there. As a matter of fact he was not, for though his body was near, he was down in the hay, and he never heard or saw the bird.

We kept watch of the fateful branch, ready to protect it if necessary, till the train moved off, and then we went home congratulating ourselves on possessing the goldfinch's precious secret, planning to spend a part of every morning in studying her ways.

"Man proposes," but many things "dispose." The next morning revealed another tragedy. The dainty nest, so laboriously built, was found a wreck, the whole of one side pulled out and hanging over the branch, while the soft cushion of silky white thistle-down, an inch thick, lay on the grass below. The culprit we could not discover, for he had left no trace. It might be a squirrel; it certainly looked like the work of his strong claws; but, on the other hand, it might be the sparrow-hawk who had made the meadow his daily hunting-ground since the mysterious disaster to the kingbird's nest had deprived us of the police services of that vigilant bird. Probably a squirrel was the culprit, for the hawk appeared only after the grass was cut, and grasshoppers and other insects were left without shelter, and he seemed to give his entire attention to the grass at the foot of the flagpole on which he always perched.

Whoever was guilty of the cruel deed, it added one more to the list of ravaged nests, and of all that we watched that summer exactly half had been broken up or destroyed.

I am happy to say that the little pair were not utterly discouraged, for a day or two later we found the provident mistress carefully drawing out of the ruin some of the material she had woven into it, and carrying it away, doubtless to add to a fresh nest. But she had this time chosen a more secluded site, that we were unable to discover. I hope she did not credit us with her disaster.

XXV. A PLUM-TREE ROMANCE

It was just after the catastrophe of the last chapter when a pair of goldfinches, whose pretty pastoral I hoped to watch, had been robbed and driven from their home in a maple-tree that the plum-tree romance began. Grieving for their sorrow as well as for my loss, I turned my steps toward the farmhouse, intending to devote part of the day to the baby crows, who were enlivening the pasture with their droll cries and droller actions. But the crow family had the pasture to themselves that morning, for in passing through the orchard, looking, as always, for indications of feathered life, I suddenly saw a new nest in the top of a plum-tree, and my spirits rose instantly when I noticed that the busy little architect, at that moment working upon it, was a goldfinch.

What an unfortunate place she had chosen, was my first thought. A young tree, a mere sapling, not more than eight feet high, close beside the regular farm road, where men, and worse, two nest-robbing boys, passed forty times a day. Would the trim little matron, now so happy in her plans, have any chance of bringing up a brood there in plain sight, where, if the roving eyes of those youngsters happened to fall upon her nest, peace would take its departure even if calamity did not overtake her?

Looking all about, to make sure that no one was in sight, I seated myself to make the acquaintance of my new neighbor. My whole study of the life in and around the plum-tree, carried on for the next two weeks, was of a spasmodic order, for I had always to take care that no spies were about before I dared even look toward the orchard. One glimpse of me in the neighborhood would have disclosed their secret to the sharp boys who knew my ways.

The little dame was bewitching in her manner, and her handsome young spouse the most devoted consort I ever saw in feathers, or out of them, I may say. Although she alone built the nest, he was her constant attendant, and they always made their appearance together. He dropped into a taller tree—an apple near by—while she, with her beak full of materials, alighted on the lowest branch of the plum, and hopped gayly from twig to twig, as though they were steps, up to the sky parlor where she had established her homestead.

Then she went busily to work to adjust the new matter, while he waited patiently during the ten or fifteen minutes she thus occupied. Sometimes he seemed to wonder what she could be about all this time, for he came and alighted beside her, staying only an instant, and then flying with the evident expectation that she would follow.

Usually, however, he remained quietly on guard till she left the nest with her joyful call, when he joined her, and away they went together, crying, "te-o-tum, te! te!" till out of sight and hearing. There was a joyousness of manner in this pair that gave a festive air to even so prosaic a performance as going for food. The source of supplies, as I soon discovered, was a bit of neglected ground between a buckwheat patch and a barn, where grass and weeds of several sorts flourished. Here each bird pulled down by its weight a stalk of meadow or other grass, and spent some time feasting upon its seeds.

But madam was a timid little soul; she reminded me constantly of some bigger folk I have known. She wanted her gay cavalier always within call, and he responded to her demands nobly, becoming more domestic than one would imagine possible for such a restless, light-hearted sprite. After the young house-mistress settled herself to her sitting, she often lifted her head above the edge of her nest, and uttered a strangely thrilling and appealing cry, which I think is only heard in the nesting-time. He always replied instantly, in tenderest tones, and came at once, sometimes from the other side of the orchard, singing as he flew, and perched in the apple-tree. If she wanted his escort to lunch, she joined him there, and after exchanging a few low remarks, they departed together.

Occasionally, however, she seemed to be merely nervous, perhaps about some other bird who she fancied might be troublesome, though, in general, neither of the pair paid the slightest attention to birds who came about, even upon their own little tree.

Often when the goldfinch came in answer to this call of his love, he flew around, at some height above the tree, in a circle of thirty or forty feet diameter, apparently to search out any enemy who might be annoying her. If he saw a bird, he drove him off, though in a perfunctory manner, as if it were done merely in deference to his lady's wishes, and not from any suspicion or jealousy. On these occasions, too, he came quite near me, stood fearless and calm, and studied me most sharply, doubtless to see if my intentions were innocent. Of course I looked as amiable and harmless as possible, and in a moment he decided that I was not dangerous, made some quiet remark to his fussy little partner, and flew away.

Sometimes this conduct did not reassure the uneasy bird, and she called again. Then he brought some tidbit in his beak, went to the edge of the nest, and fed her. Then she was pacified; but do not mistake her, it was not hunger that prompted her actions; when she was hungry, she openly left her nest and went for food. It was, as I am convinced, the longing desire to know that he was near her, that he was still anxious to serve her, that he had not forgotten her in her long absence from his side. This may sound a little fanciful to one who has not studied birds closely, but she was so "human" in all her actions that I feel justified in judging of her motives exactly as I should judge had she measured five feet instead of five inches, and worn silk instead of feathers.

The goldfinch need not have worried about her mate, for he spent most of his time within a few feet of her, and more absolutely loyal one could not be. His most common perch was a neighboring tree, though in a heavy beating rain he frequently crouched on the lowest branch of the plum itself. Now and then he rested on a pile of boards beside the farm road already spoken of, and again he took his post on a very tall ash, with only a few limbs at the top, where his body looked like a dot against the blue, and he could oversee the whole country around. Wherever he might be, he sat all puffed out, silent and motionless, evidently just waiting. Sometimes he took occasion to plume himself very carefully, oftener he did nothing, but held himself in readiness to answer any call from the plum-tree, and to accompany the sitter out to dinner.

This bird was quite an enchanting singer. During courtship, and while his mate was sitting there, he often poured out a song that was nothing less than an ecstasy. It was delivered on the wing, and not in his usual wave-like manner of flight, but sailing slowly around and around, very much as a bobolink does, singing rapturously, without pause or break. The quality of the music, too, was strikingly like bobolink notes, and the whole performance was exquisite.

The little sitter soon became accustomed to my presence. When out of her nest, she sometimes came to the tree over my head, and answered when I spoke to her. In this way we carried on quite a long conversation, I imitating, so far as I was able, her own charming "sweet," and she replying in varied utterances, which, alas! were Greek to me.

I longed to watch the loving pair through their nesting; to see their rapture over their nestlings, their tender care and training, and the first flight of the goldfinch babies. But the inexorable task-master of us all, who proverbially "waits for no man," hurried off these last precious days of July with painful eagerness, and thrust before me the first of August, with the hot and dusty journey set down for that day, long be-fore I was ready for it.

So I did not see the end of their love and labor myself, but the bird's wisdom in the selection of a site for her nursery was proved to be greater than mine, who had ventured to criticize her, by the fact that the nest, as I have been assured, escaped the young eyes of the neighborhood, and turned out its full complement of birdlings to add to next summer's beauty and song.

XXVI. SOLITARY THE THRUSH

"Solitary the thrush,
The hermit, withdrawn to himself,
Sings by himself a song."

Thus says the poet, with no less truth than beauty. No description could better express the spirit of the bird, the retiring habit and the love of quiet for which not alone the hermit, but the three famous singers of the thrush family are remarkable. We should indeed be shocked were it otherwise, for there is an indefinable quality in the tones of this trio, the hermit, wood, and tawny, that stirs the soul to its depths, and one can hardly conceive of them as mingling their notes with other singers, or becoming in any way familiar. In this peculiar power no bird-voice in our part of the world can compare with theirs.

The brown thrush ranks high as a musician, the mockingbird leads the world, in the opinion of its lovers, and the winter wren thrills one to the heart. Yet no bird song so moves the spirit, no other—it seems to me—so intoxicates its hearer with rapture, as the solemn chant of "the hermit withdrawn to himself."

"Whenever a man hears it," says our devoted lover of Nature, Thoreau, "he is young, and Nature is in her spring; wherever he hears it there is a new world, and the gates of heaven are not shut against him."

One might quote pages of rhapsody from poets and prose writers, yet to him who has not drunk of the enchantment, they would be but words; they would touch no chord that had not already been thrilled by the marvelous strain itself.

My first acquaintance in the beautiful family was the wood-thrush, and the study of his charms of voice and character filled me with love for the whole bird tribe. He frequented the places I also preferred, the quiet nooks and out of the way corners of a large city park. At that time I thought no bird note on earth could equal his; but a year or two later, on the shore of Lake George, I fell under the magical sway of another voice, whose few notes were exceedingly simple in arrangement, but full of the strangely thrilling power characteristic of the thrush family.

Four years passed, at first in search of the owner of the "wandering voice" that had bewitched me, and when I had found it to be the tawny thrush or veery, in study of the attractive singer himself, which made me an enthusiastic lover of him also. But the "shy and hidden" bird, the hermit, enthroned by those who know him far above the others, I had rarely seen and never clearly heard. Far-off snatches I had gathered, and a few of the louder notes had reached me from distant woods, or from farther up the mountain side; but I had never been satisfied.

There appeared almost a fatality about my hearing this bird. No matter how common his song in the neighborhood, no sooner did I go there than he retired to the secluded recesses of his choice. He always had "just been singing," but had mysteriously stopped. My search was much longer than, and quite as disappointing as Mr. Burroughs's search through English lanes for a singing nightingale.

Last spring one of the strongest attractions that drew me to a lovely spot in Northern New York was the assurance that the hermit was a constant visitor. I went, and the same old story met me. Before this year the hermit had always been with them. The song of the veery was my morning and evening inspiration, but his shy brother had apparently taken his departure for parts unknown.

"We will go to Sunset Hill," said my friend. "We always hear them there at sunset."

That evening after an early tea, we started for the promised land. The single-file procession through the charming wood paths consisted of our host as protector on the return in the dark, the big dog—his mistress's body-guard—his mistress, an enthusiastic bird-lover, and myself.

The road was all the way through the woods, then lovely with the glow of the western sun, which reached far under the branches, gilded the trunks of the trees, and made a fresh picture at every turn. At the further side of the woods was a grass-covered hill which we ascended, eager to treat our eyes to the sunset, and our ears to the hermit songs. The sun went down serenely, without a cloud to reflect his glory, but the whole pleasant country at our feet was illuminated by his parting rays.

And hark! a hermit began "air-o-ee!" Instantly everything else was forgotten, although the bird was far away.

"He will come nearer," whispered my comrade, and we waited in silence. Several singers were within hearing, but all at a tantalizing remoteness that allowed us to hear the louder notes, and constantly to realize what we were losing.

We lingered, loath to abandon hope, till the deepening shadows reminded us of the woods to be passed through; but no bird came nearer than that maddening distance. In despair we turned our faces homeward at last; several times on the way we paused, lured by an ecstatic note, but every one too far off to be completely heard.

In our quiet walk back through the dark woods I accepted my evident fate, that I was not to be blessed with hermit music this season; but I made a private resolve to find next year a "hermit neighborhood," where birds should be warranted to sing, if I had to take a tent and camp out in a swamp.

June passed away in delightful bird-study, and July followed quickly. Nests and songs in plenty rewarded our search. Every day had been full. Nothing had been wanting to fill our cup of content, except the longed-for song of the hermit; and I had been so absorbed I had almost ceased to regret it.

With the last days of July everything was changed about us. The world was full of bird babies. Infant voices rang out from every tangle; flutters of baby wings stirred every bush; the woods echoed to anxious "pips," and "smacks," and "quits," of uneasy parents working for dear life. We had been so occupied with our study of these charming youngsters, that we bethought ourselves, only as one after another strange warbler appeared upon the scene, that migrating time had arrived, the wonderful procession to the summer-land had begun.

This, alas! I could not stay to see. And if one must go, it were better to take leave before getting entangled in the toils of the warblers, to be driven wild by the numberless shades of yellow and olive, to go frantic over stripes and spots, and bars, and to wear out patience and the Manual, trying to discover what particular combination of Latin syllables scientists have bestowed upon this or that flitting atom in feathers. Before the student is out of bed, a new warbler-note will distract her; in the twilight some tiny bird will fly over her head with an unfamiliar twitter; each and every one will rouse her to eager desire to see it, to name it.

Why have we such a rage for labeling and cataloguing the beautiful things of Nature? Why can I not delight in a bird or flower, knowing it by what it is to me, without longing to know what it has been to some other person? What pleasure can it afford to one not making a scientific study of birds to see such names as "the blue and yellow-throated warbler," "the chestnut-headed golden warbler," "the yellow-bellied, red-poll warbler," attached to the smallest and daintiest beauties of the woods?

Musing upon this and other mysteries, I followed my friend up the familiar paths one day, looking for some young birds whose strange cries we had noted. It was a gray morning, and all the tree trunks were grim and dark, with no variety in coloring. The sounds we were following led us through some unused roads entirely grown up with jewel-weed, part of it five feet high, and thickly hung with the yellow flower from which it takes its name.

It had rained in the night, and every leaf was adorned with minute drops like gems. We parted the stems carefully and passed through, though it seemed to us like wading in deep water, and, in spite of our caution, we were well sprinkled from the dripping leaves. Just as we stepped out of our green sea, the low calls we were trying to locate ceased. We walked slowly on until we were attracted by a rustling in the dry leaves, and then we turned to see two young thrushes foraging about in silence by themselves. They were not very shy, but looked at us with innocent baby eyes as we drew near and examined them. We saw the color and the markings and the peculiar movement of the tail characteristic of the hermit. There could be no doubt that these were hermit babies. We were delighted to see them. I never feel that I know a bird family till I have seen the young. But my pleasure was sadly marred by the reflection that where there were babies must have been a nest and a singer, and we had not heard his voice.

The last Sunday of my stay came, all too soon. It was a glorious day, and, as usual, the two bird-lovers turned their steps toward the woods. Everything seemed at rest and silent. We paused a while in a part of the forest in which we had seen some strange phases of bird life, and had christened the "Bewitched Corner." A gentle breeze set all the leaves to fluttering; far off a woodpecker drummed his salute to his fellows; beyond the trees we could hear the indigo bird singing; but nothing about us was stirring. The wood-pewee was unheard, and even the vireo seemed to have finished his endless song and gone his way.

We passed on a few rods to a favorite resting place of our daily rounds, where my comrade always liked to stretch herself upon the big bole of a fallen tree in the broad sunshine, and I to seat myself at the foot of another tree in the shade. It was a spot

"where hours went their way
As softly as sweet dreams go down the night."

As we approached this place a sound reached us that struck us dumb; it was a hermit thrush not far off. Silently we stole up the gentle hill and seated ourselves.

"At last! at last!" I cried in my heart, as I leaned back against my tree to listen.

Then the glorious anthem began again; it rose and swelled upon the air; it filled the woods—"And up by mystical chords of song
The soul was lifted from care and pain."

Though not in sight, the bird was quite near, so that we heard every note, so enchanting! so inimitable! For ten or fifteen minutes he poured out the melody, while our hearts fairly stood still. Then he stopped, and we heard the thrush "chuck" and the hermit call, which is different from other thrushes, being something between a squawk and a mew. Whether this were his conversation with his mate we could only guess, for we dared not move, hardly indeed to breathe.

After a pause the bird began again, and for one perfect hour we sat there motionless, entranced, and took our fill of his matchless rhapsody. I longed inexpressibly to see the enchanter, though I dared not stir for fear of startling him. Perhaps my urgent desire drew him; at any rate he came at last within sight, stood a few minutes on the low branch of a tree and looked at me, lifting and dropping his expressive tail as he did so. Two or three low, rich notes bubbled out, as if he had half a mind to sing to me; but he thought better of it and dived off the branch into the bushes. We rose to go.

"This only was lacking," I said. "This crowns my summer. I ask no more, and tomorrow I go."

INDEX

Idle Winter Press

www.ingramcontent.com/pod-product-compliance
Lightning Source LLC
Chambersburg PA
CBHW071019280326
41935CB00011B/1419
* 9 7 8 0 6 9 2 3 8 1 2 2 9 *